Making Sense of Data

THE WILEY BICENTENNIAL–KNOWLEDGE FOR GENERATIONS

Each generation has its unique needs and aspirations. When Charles Wiley first opened his small printing shop in lower Manhattan in 1807, it was a generation of boundless potential searching for an identity. And we were there, helping to define a new American literary tradition. Over half a century later, in the midst of the Second Industrial Revolution, it was a generation focused on building the future. Once again, we were there, supplying the critical scientific, technical, and engineering knowledge that helped frame the world. Throughout the 20th Century, and into the new millennium, nations began to reach out beyond their own borders and a new international community was born. Wiley was there, expanding its operations around the world to enable a global exchange of ideas, opinions, and know-how.

For 200 years, Wiley has been an integral part of each generation's journey, enabling the flow of information and understanding necessary to meet their needs and fulfill their aspirations. Today, bold new technologies are changing the way we live and learn. Wiley will be there, providing you the must-have knowledge you need to imagine new worlds, new possibilities, and new opportunities.

Generations come and go, but you can always count on Wiley to provide you the knowledge you need, when and where you need it!

WILLIAM J. PESCE
PRESIDENT AND CHIEF EXECUTIVE OFFICER

PETER BOOTH WILEY
CHAIRMAN OF THE BOARD

Making Sense of Data

A Practical Guide to Exploratory Data Analysis and Data Mining

Glenn J. Myatt

BICENTENNIAL
1807
WILEY
2007
BICENTENNIAL

WILEY-INTERSCIENCE
A JOHN WILEY & SONS, INC., PUBLICATION

Library of Congress Cataloging-in-Publication Data

ISBN-13: 978-0-470-07471-8
ISBN-10: 0-470-07471-X

10 9 8 7 6 5

Contents

Preface

Almost every field of study is generating an unprecedented amount of data. Retail companies collect data on every sales transaction, organizations log each click made on their web sites, and biologists generate millions of pieces of information related to genes daily. The volume of data being generated is leading to *information overload* and the ability to make sense of all this data is becoming increasingly important. It requires an understanding of exploratory data analysis and data mining as well as an appreciation of the subject matter, business processes, software deployment, project management methods, change management issues, and so on.

The purpose of this book is to describe a practical approach for making sense out of data. A step-by-step process is introduced that is designed to help you avoid some of the common pitfalls associated with complex data analysis or data mining projects. It covers some of the more common tasks relating to the analysis of data including (1) how to summarize and interpret the data, (2) how to identify nontrivial facts, patterns, and relationships in the data, and (3) how to make predictions from the data.

The process starts by understanding what business problems you are trying to solve, what data will be used and how, who will use the information generated and how will it be delivered to them. A plan should be developed that includes this problem definition and outlines how the project is to be implemented. Specific and measurable success criteria should be defined and the project evaluated against them.

The relevance and the quality of the data will directly impact the accuracy of the results. In an ideal situation, the data has been carefully collected to answer the specific questions defined at the start of the project. Practically, you are often dealing with data generated for an entirely different purpose. In this situation, it will be necessary to prepare the data to answer the new questions. This is often one of the most time-consuming parts of the data mining process, and numerous issues need to be thought through.

Once the data has been collected and prepared, it is now ready for analysis. What methods you use to analyze the data are dependent on many factors including the problem definition and the type of data that has been collected. There may be many methods that could potentially solve your problem and you may not know which one works best until you have experimented with the different alternatives. Throughout the technical sections, issues relating to when you would apply the different methods along with how you could optimize the results are discussed.

Once you have performed an analysis, it now needs to be delivered to your target audience. This could be as simple as issuing a report. Alternatively, the delivery may involve implementing and deploying new software. In addition to any technical challenges, the solution could change the way its intended audience operates on a daily basis, which may need to be managed. It will be important to understand how well the solution implemented in the field actually solves the original business problem.

Any project is ideally implemented by an interdisciplinary team, involving subject matter experts, business analysts, statisticians, IT professionals, project managers, and data mining experts. This book is aimed at the entire interdisciplinary team and addresses issues and technical solutions relating to data analysis or data mining projects. The book could also serve as an introductory textbook for students of any discipline, both undergraduate and graduate, who wish to understand exploratory data analysis and data mining processes and methods.

The book covers a series of topics relating to the process of making sense of data, including

- Problem definitions
- Data preparation
- Data visualization
- Statistics
- Grouping methods
- Predictive modeling
- Deployment issues
- Applications

The book is focused on practical approaches and contains information on how the techniques operate as well as suggestions for when and how to use the different methods. Each chapter includes a further reading section that highlights additional books and online resources that provide background and other information. At the end of selected chapters are a set of exercises designed to help in understanding the respective chapter's materials.

Accompanying this book is a web site (http://www.makingsenseofdata.com/) containing additional resources including software, data sets, and tutorials to help in understanding how to implement the topics covered in this book.

In putting this book together, I would like to thank the following individuals for their considerable help: Paul Blower, Vinod Chandnani, Wayne Johnson, and Jon Spokes. I would also like to thank all those involved in the review process for the book. Finally, I would like to thank the staff at John Wiley & Sons, particularly Susanne Steitz, for all their help and support throughout the entire project.

Chapter 1

Introduction

1.1 OVERVIEW

Disciplines as diverse as biology, economics, engineering, and marketing measure, gather and store data primarily in electronic databases. For example, retail companies store information on sales transactions, insurance companies keep track of insurance claims, and meteorological organizations measure and collect data concerning weather conditions. Timely and well-founded decisions need to be made using the information collected. These decisions will be used to maximize sales, improve research and development projects and trim costs. Retail companies must be able to understand what products in which stores are performing well, insurance companies need to identify activities that lead to fraudulent claims, and meteorological organizations attempt to predict future weather conditions. The process of taking the raw data and converting it into meaningful information necessary to make decisions is the focus of this book.

It is practically impossible to make sense out of data sets containing more than a handful of data points without the help of computer programs. Many free and commercial software programs exist to sift through data, such as spreadsheets, data visualization software, statistical packages, OLAP (On-Line Analytical Processing) applications, and data mining tools. Deciding what software to use is just one of the questions that must be answered. In fact, there are many issues that should be thought through in any exploratory data analysis/data mining project. Following a predefined process will ensure that issues are addressed and appropriate steps are taken.

Any exploratory data analysis/data mining project should include the following steps:

1. **Problem definition:** The problem to be solved along with the projected deliverables should be clearly defined, an appropriate team should be put together, and a plan generated for executing the analysis.

2. **Data preparation:** Prior to starting any data analysis or data mining project, the data should be collected, characterized, cleaned, transformed, and partitioned into an appropriate form for processing further.

Making Sense of Data: A Practical Guide to Exploratory Data Analysis and Data Mining,
By Glenn J. Myatt
Copyright © 2007 John Wiley & Sons, Inc.

3. **Implementation of the analysis:** On the basis of the information from steps 1 and 2, appropriate analysis techniques should be selected, and often these methods need to be optimized.

4. **Deployment of results:** The results from step 3 should be communicated and/or deployed into a preexisting process.

Although it is usual to follow the order described, there will be some inter-actions between the different steps. For example, it may be necessary to return to the data preparation step while implementing the data analysis in order to make modifications based on what is being learnt. The remainder of this chapter summarizes these steps and the rest of the book outlines how to execute each of these steps.

1.2   PROBLEM DEFINITION

The first step is to define the business or scientific problem to be solved and to understand how it will be addressed by the data analysis/data mining project. This step is essential because it will create a focused plan to execute, it will ensure that issues important to the final solution are taken into account, and it will set correct expectations for those both working on the project and having a stake in the project's results. A project will often need the input of many individuals including a specialist in data analysis/data mining, an expert with knowledge of the business problems or subject matter, information technology (IT) support as well as users of the results. The plan should define a timetable for the project as well as providing a comparison of the cost of the project against the potential benefits of a successful deployment.

1.3   DATA PREPARATION

In many projects, getting the data ready for analysis is the most time-consuming step in the process. Pulling the data together from potentially many different sources can introduce difficulties. In situations where the data has been collected for a different purpose, the data will need to be transformed into an appropriate form for analysis. During this part of the project, a thorough familiarity with the data should be established.

1.4   IMPLEMENTATION OF THE ANALYSIS

Any task that involves making decisions from data almost always falls into one of the following categories:

- **Summarizing the data:** *Summarization is a process in which the data is reduced for interpretation without sacrificing any important information.* Summaries can be developed for the data as a whole or any portion of the data. For example, a retail company that collected data on its transactions could develop summaries of the total sales transactions. In addition, the company could also generate summaries of transactions by products or stores.

- **Finding hidden relationships:** *This refers to the identification of important facts, relationships, anomalies or trends in the data, which are not obvious from a summary alone.* To discover this information will involve looking at the data from many angles. For example, a retail company may want to understand customer profiles and other facts that lead to the purchase of certain product lines.

- **Making predictions:** *Prediction is the process where an estimate is calculated for something that is unknown.* For example, a retail company may want to predict, using historical data, the sort of products that specific consumers may be interested in.

There is a great deal of interplay between these three tasks. For example, it is important to summarize the data before making predictions or finding hidden relationships. Understanding any hidden relationships between different items in the data can help in generating predictions. Summaries of the data can also be useful in presenting prediction results or understanding hidden relationships identified. This overlap between the different tasks is highlighted in the Venn diagram in Figure 1.1.

Exploratory data analysis and data mining covers a broad set of techniques for summarizing the data, finding hidden relationships, and making predictions. Some of the methods commonly used include

- **Summary tables:** The raw information can be summarized in multiple ways and presented in tables.

- **Graphs:** Presenting the data graphically allows the eye to visually identify trends and relationships.

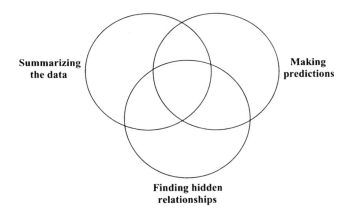

Figure 1.1.   Data analysis tasks

- **Descriptive statistics:** These are descriptions that summarize information about a particular data column, such as the average value or the extreme values.

- **Inferential statistics:** Methods that allow claims to be made concerning the data with confidence.

- **Correlation statistics:** Statistics that quantify relationships within the data.

- **Searching:** Asking specific questions concerning the data can be useful if you understand the conclusion you are trying to reach or if you wish to quantify any conclusion with more information.

- **Grouping:** Methods for organizing a data set into smaller groups that potentially answer questions.

- **Mathematical models:** A mathematical equation or process that can make predictions.

The three tasks outlined at the start of this section (summarizing the data, finding hidden relationships, and making predictions) are shown in Figure 1.2 with a circle for each task. The different methods for accomplishing these tasks are also positioned on the Venn diagram. The diagram illustrates the overlap between the various tasks and the methods that can be used to accomplish them. The position of the methods is related to how they are often used to address the various tasks.

Graphs, summary tables, descriptive statistics, and inferential statistics are the main methods used to summarize data. They offer multiple ways of describing the data and help us to understand the relative importance of different portions of the data. These methods are also useful for characterizing the data prior to developing predictive models or finding hidden relationships. Grouping observations can be useful in teasing out hidden trends or anomalies in the data. It is also useful for characterizing the data prior to building predictive models. Statistics are used

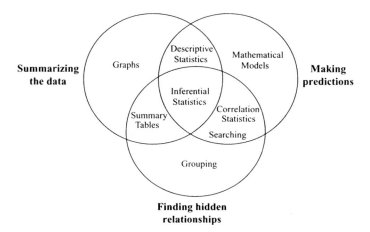

Figure 1.2.   Data analysis tasks and methods

throughout, for example, correlation statistics can be used to prioritize what data to use in building a mathematical model and inferential statistics can be useful when validating trends identified from grouping the data. Creating mathematical models underpins the task of prediction; however, other techniques such as grouping can help in preparing the data set for modeling as well as helping to explain why certain predictions were made.

All methods outlined in this section have multiple uses in any data analysis or data mining project, and they all have strengths and weaknesses. On the basis of issues important to the project as well as other practical considerations, it is necessary to select a set of methods to apply to the problem under consideration. Once selected, these methods should be appropriately optimized to improve the quality of the results generated.

1.5 DEPLOYMENT OF THE RESULTS

There are many ways to deploy the results of a data analysis or data mining project. Having analyzed the data, a static report to management or to the customer of the analysis is one option. Where the project resulted in the generation of predictive models to use on an ongoing basis, these models could be deployed as standalone applications or integrated with other softwares such as spreadsheets or web pages. It is in the deployment step that the analysis is translated into a benefit to the business, and hence this step should be carefully planned.

1.6 BOOK OUTLINE

This book follows the four steps outlined in this chapter:

1. **Problem definition:** A discussion of the definition step is provided in Chapter 2 along with a case study outlining a hypothetical project plan. The chapter outlines the following steps: (1) define the objectives, (2) define the deliverables, (3) define roles and responsibilities, (4) assess the current situation, (5) define the timetable, and (6) perform a cost/benefit analysis.

2. **Data preparation:** Chapter 3 outlines many issues and methods for preparing the data prior to analysis. It describes the different sources of data. The chapter outlines the following steps: (1) create the data tables, (2) characterize the data, (3) clean the data, (4) remove unnecessary data, (5) transform the data, and (6) divide the data into portions when needed.

3. **Implementation of the analysis:** Chapter 4 provides a discussion of how summary tables and graphs can be used for communicating information about the data. Chapter 5 reviews a series of useful statistical approaches to summarizing the data and relationships within the data as well as making statements about the data with confidence. It covers the following topics: descriptive statistics, confidence intervals, hypothesis tests, the chi-square test, one-way analysis of variance, and correlation analysis. Chapter 6 describes a

series of methods for grouping data including clustering, associative rules, and decision trees. Chapter 7 outlines the process and methods to be used in building predictive models. In addition, the chapter covers a series of methods including simple regression, k-nearest neighbors, classification and regression trees, and neural networks.

4. **Deployment of results:** Chapter 8 reviews some of the issues around deploying any results from data analysis and data mining projects including planning and executing deployment, measuring and monitoring the solution's performance, and reviewing the entire project. A series of common deployment scenarios are presented. Chapter 9 concludes the book with a review of the whole process, a case study, and a discussion of data analysis

Table 1.1.   Summary of project steps

Steps	Description
1. Problem definition	Define • Objectives • Deliverables • Roles and responsibilities • Current situation • Timeline • Costs and benefits
2. Data preparation	Prepare and become familiar with the data: • Pull together data table • Categorize the data • Clean the data • Remove unnecessary data • Transform the data • Partition the data
3. Implementation of the analysis	Three major tasks are • Summarizing the data • Finding hidden relationships • Making prediction Select appropriate methods and design multiple experiments to optimize the results. Methods include • Summary tables • Graphs • Descriptive statistics • Inferential statistics • Correlation statistics • Searching • Grouping • Mathematical models
4. Deployment	• Plan and execute deployment based on the definition in step 1 • Measure and monitor performance • Review the project

and data mining issues associated with common applications. Exercises are included at the end of selected chapters to assist in understanding the material.

This book uses a series of data sets to illustrate the concepts from Newman (1998). The Auto-Mpg Database is used throughout to compare how the different approaches view the same data set. In addition, the following data sets are used in the book: Abalone Database, Adult Database, and the Pima Indians Diabetes Database.

1.7 SUMMARY

The four steps in any data analysis or data mining project are summarized in Table 1.1.

1.8 FURTHER READING

The CRISP-DM project (CRoss Industry Standard Process for Data Mining) has published a data mining process and describes details concerning data mining stages and relationships between the stages. It is available on the web at: http://www.crisp-dm.org/

SEMMA (Sample, Explore, Modify, Model, Assess) describes a series of core tasks for model development in the SAS[R] Enterprise Miner[TM] software and a description can be found at: http://www.sas.com/technologies/analytics/datamining/miner/semma.html

Chapter 2

Definition

2.1 OVERVIEW

This chapter describes a series of issues that should be considered at the start of any data analysis or data mining project. It is important to define the problem in sufficient detail, in terms of both how the questions are to be answered and how the solutions will be delivered. On the basis of this information, a cross-disciplinary team should be put together to implement these objectives. A plan should outline the objectives and deliverables along with a timeline and budget to accomplish the project. This budget can form the basis for a cost/benefit analysis, linking the total cost of the project to potential savings or increased revenues. The following sections explore issues concerning the problem definition step.

2.2 OBJECTIVES

It is critical to spend time defining how the project will impact specific business objectives. This assessment is one of the key factors to achieving a successful data analysis/data mining project. Any technical implementation details are secondary to the definition of the business objective. Success criteria for the project should be defined. These criteria should be specific and measurable as well as related to the business objective. For example, the project should increase revenue or reduce costs by a specific amount.

A broad description of the project is useful as a headline. However, this description should be divided into smaller problems that ultimately solve the broader issue. For example, a general problem may be defined as: "Make recommendations to improve sales on the web site." This question may be further broken down into questions that can be answered using the data such as:

1. Identify categories of web site users (on the basis of demographic information) that are more likely to purchase from the web site.
2. Categorize users of the web site on the basis of usage information.

Making Sense of Data: A Practical Guide to Exploratory Data Analysis and Data Mining,
By Glenn J. Myatt
Copyright © 2007 John Wiley & Sons, Inc.

3. Determine if there are any relationships between buying patterns and web site usage patterns.

All those working on the project as well as other interested parties should have a clear understanding of what problems are to be addressed. It should also be possible to answer each problem using the data. To make this assessment, it is important to understand what the collection of all possible observations that would answer the question would look like or *population*. For example, when the question is how America will vote in the upcoming presidential election, then the entire population is all eligible American voters. Any data to be used in the project should be representative of the population. If the problems cannot be answered with the available data, a plan describing how this data will be acquired should be developed.

2.3 DELIVERABLES

It is also important to identify the deliverables of the project. Will the solution be a report, a computer program to be used for making predictions, a new workflow or a set of business rules? Defining all deliverables will provide the correct expectations for all those working on the project as well as any project stakeholders, such as the management who is sponsoring the project.

When developing predictive models, it is useful to understand any required level of accuracy. This will help prioritize the types of approaches to consider during implementation as well as focus the project on aspects that are critical to its success. For example, it is not worthwhile spending months developing a predictive model that is 95% accurate when an 85% accurate model that could have been developed in days would have solved the business problem. This time may be better devoted to other aspects that influence the ultimate success of the project. The accuracy of the model can often relate directly to the business objective. For example, a credit card company may be suffering due to customers moving their accounts to other companies. The company may have a business objective of reducing this turnover rate by 10%. They know that if they are able to identify a customer that is likely to abandon their services, they have an opportunity of targeting and retaining this customer. The company decides to build a prediction model to identify these customers. The level of accuracy of the prediction, therefore, has to be such that the company can reduce the turnover by the desired amount.

It is also important to understand the consequences of answering questions incorrectly. For example, when predicting tornadoes, there are two possible scenarios: (1) incorrectly predicting a tornado and (2) incorrectly predicting no tornado. The consequence of scenario (2) is that a tornado hits with no warning. Affected neighborhoods and emergency crews would not be prepared for potentially catastrophic consequences. The consequence of scenario (1) is less dramatic with only a minor inconvenience to neighborhoods and emergency services since they prepared for a tornado that did not hit. It is usual to relate business consequences to the quality of prediction according to these two scenarios.

One possible deliverable is a software application, such as a web-based data mining application that suggests alternative products to customers while they are browsing an online store. The time to generate an answer is dependent, to a large degree, on the data mining approach adopted. If the speed of the computation is a factor, it must be singled out as a requirement for the final solution. In the online shopping example, the solution must generate these items rapidly (within a few seconds) or the customer will become frustrated and shop elsewhere.

In many situations, the time to create a model can have an impact on the success of the project. For example, a company developing a new product may wish to use a predictive model to prioritize potential products for testing. The new product is being developed as a result of competitive intelligence indicating that another company is developing a similar product. The company that is first to the market will have a significant advantage. Therefore, the time to generate the model may be an important factor since there is only a window of opportunity to influence the project. If the model takes too long to develop, the company may decide to spend considerable resources actually testing the alternatives as opposed to making use of any models generated.

There are a number of deployment issues that may need to be considered during the implementation phase. A solution may involve changing business processes. For example, a solution that requires the development of predictive models to be used by associates in the field may change the work practices of these individuals. These associates may even resist this change. Involving the end-users in the project may facilitate acceptance. In addition, the users may require that all results are appropriately explained and linked to the data from which the results were generated, in order to trust the results.

Any plan should define these and other issues important to the project as these issues have implications as to the sorts of methods that can be adopted in the implementation step.

2.4   ROLES AND RESPONSIBILITIES

It is helpful to consider the following roles that are important in any project.

- **Project leader:** Someone who is responsible for putting together a plan and ensuring the plan is executed.
- **Subject matter experts and/or business analysts:** Individuals who have specific knowledge of the subject matter or business problems including (1) how the data was collected, (2) what the data values mean, (3) the level of accuracy of the data, (4) how to interpret the results of the analysis, and (5) the business issues being addressed by the project.
- **Data analysis/data mining expert:** Someone who is familiar with statistics, data analysis methods and data mining approaches as well as issues of data preparation.

- **IT expert:** A person or persons with expertise in pulling data sets together (e.g., accessing databases, joining tables, pivoting tables, etc.) as well as knowledge of software and hardware issues important for the implementation and deployment steps.

- **Consumer:** Someone who will ultimately use the information derived from the data in making decisions, either as a one-off analysis or on a routine basis.

A single member of the team may take on multiple roles such as an individual may take on the role of project leader and data analysis/data mining expert. Another scenario is where multiple persons are responsible for a single role, for example, a team may include multiple subject matter experts, where one individual has knowledge of how the data was measured and another individual has knowledge of how the data can be interpreted. Other individuals, such as the project sponsor, who have an interest in the project should be brought in as interested parties. For example, representatives from the finance group may be involved in a project where the solution is a change to a business process with important financial implications.

Cross-disciplinary teams solve complex problems by looking at the data from different perspectives and should ideally work on these types of projects. Different individuals will play active roles at different times. It is desirable to involve all parties in the definition step. The IT expert has an important role in the data preparation step to pull the data together in a form that can be processed. The data analysis/data mining expert and the subject matter expert/business analyst should also be working closely in the preparation step to clean and categorize the data. The data analysis/data mining expert should be primarily responsible for transforming the data into an appropriate form for analysis. The third implementation step is primarily the responsibility of the data analysis/data mining expert with input from the subject matter expert/business analyst. Also, the IT expert can provide a valuable hardware and software support role throughout the project.

With cross-disciplinary teams, communication challenges may arise from time-to-time. A useful way of facilitating communication is to define and share glossaries defining terms familiar to the subject matter experts or to the data analysis/data mining experts. Team meetings to share information are also essential for communication purposes.

2.5 PROJECT PLAN

The extent of any project plan depends on the size and scope of the project. However, it is always a good idea to put together a plan. It should define the problem, the proposed deliverables along with the team who will execute the analysis, as described above. In addition, the current situation should be assessed. For example, are there constraints on the personnel that can work on the project or are there hardware and software limitations that need to be taken into account? The sources and locations of the data to be used should be identified. Any issues, such as privacy or legal issues, related to using the data should be listed. For example, a data set containing personal information on customers' shopping habits could be used in a data mining project. However, information that relates directly to any individual cannot be presented as results.

A timetable of events should be put together that includes the preparation, implementation, and deployment steps. It is very important to spend the appropriate amount of time getting the data ready for analysis, since the quality of the data ultimately determines the quality of the analysis results. Often this step is the most time-consuming, with many unexpected problems with the data coming to the surface. On the basis of an initial evaluation of the problem, a preliminary implementation plan should be put together. Time should be set aside for iteration of activities as the solution is optimized. The resources needed in the deployment step are dependent on how the deliverables were previously defined. In the case where the solution is a report, the whole team should be involved in writing the report. Where the solution is new software to be deployed, then a software development and deployment plan should be put together, involving a managed roll-out of the solution.

Time should be built into the timetable for reviews after each step. At the end of the project, a valuable exercise is to spend time evaluating what worked and what did not work during the project, providing insights for future projects. It is also likely that the progress will not always follow the predefined sequence of events, moving between stages of the process from time-to-time. There may be a number of high-risk steps in the process, and these should be identified and contingencies built into the plan. Generating a budget based on the plan could be used, alongside the business success criteria, to understanding the cost/benefits for the project. To measure the success of the project, time should be set aside to evaluate if the solutions meets the business goals during deployment. It may also be important to monitor the solution over a period of time.

2.6 CASE STUDY

2.6.1 Overview

The following is a hypothetical case study involving a financial company's credit card division that wishes to reduce the number of customers switching services. To achieve this, marketing management decides to initiate a data mining project to help predict which customers are likely to switch services. These customers will be targeted with an aggressive direct marketing campaign. The following is a summarized plan for accomplishing this objective.

2.6.2 Problem

The credit card division would like to increase revenues by $2,000,000 per year by retaining more customers. This goal could be achieved if the division could predict with a 70% accuracy rate which customers are going to change services. The 70% accuracy number is based on a financial model described in a separate report. In addition, factors that are likely to lead to customers changing service will be useful in formulating future marketing plans.

To accomplish this business objective, a data mining project is established to solve the following problems:

1. Create a prediction model to forecast which customers are likely to change credit cards.
2. Find hidden facts, relationships, and patterns that customers exhibit prior to switching credit cards.

The target population is all credit card customers.

2.6.3 Deliverables

There will be two deliverables:

1. Software to predict customers likely to change credit cards.
2. A report describing factors that contribute to customers changing credit cards.

The prediction is to be used within the sales department by associates who market to at risk customers. No explanation of the results is required. The consequence of missing a customer that changes service is significantly greater than mistakenly identifying a customer that is not considering changing services. It should be possible to rank customers based on most-to-least likely to switch credit cards.

2.6.4 Roles and Responsibilities

The following individuals will work directly on the project:

- Pam (Project leader and business analyst)
- Lee (IT expert)
- Tony (Data mining consultant)

The following will serve on the team as interested parties, as they represent the customers of the solution:

- Jeff (Marketing manager and project sponsor)
- Kim (Sales associate)

2.6.5 Current Situation

A number of databases are available for use with this project: (1) a credit card transaction database and (2) a customer profile database containing information on demographics, credit ratings, as well as wealth indicators. These databases are located in the IT department.

2.6.6   Timetable and Budget

Prior to starting the project, a kick-off meeting will take place where the goals will be fine-tuned and any cross-team education will take place.

The following outlines the steps required for this project:

1. *Preparation*: Access, characterize and prepare the data sets for analysis and develop an appreciation of the data content.

2. *Implementation*: A variety of data analysis/data mining methods will be explored and the most promising optimized. The analysis will focus on creating a model to predict customers likely to switch credit cards with an accuracy greater than 70% and the discovery of factors contributing to customers changing cards.

3. *Deployment*: A two phase roll-out of the solution is planned. Phase one will assess whether the solution translates into the business objectives. In this phase, the sales department responsible for targeting at risk customers will be divided into two random groups. The first group will use the prediction models to prioritize customers. The second group will be assigned a random ranking of customers. The sales associates will not know whether they are using the prediction model or not. Differences in terms of retention of customers will be compared between the two groups. This study will determine whether the accuracy of the model translates into meeting the business objectives. When phase one has been successfully completed, a roll-out of the solution will take place and changes will be made to the business processes.

A meeting will be held after each stage of the process with the entire group to review what has been accomplished and agree on a plan for the next stage.

There are a number of risks and contingencies that need to be built into the plan. If the model does not have a required accuracy of 70%, any deployment will not result in the desired revenue goals. In this situation, the project should be reevaluated. In the deployment phase, if the projected revenue estimates from the double blind test does not meet the revenue goals then the project should be reevaluated at this point.

Figure 2.1 shows a timetable of events and a summarized budget for the project.

2.6.7   Cost/Benefit Analysis

The cost of the project of $35,500 is substantially less than the projected saving of $2,000,000. A successfully delivered project would have a substantial return on investment.

2.7   SUMMARY

Table 2.1 summarizes the problem definition step.

Steps	Tasks	Parties involved	Deliverables	Apr	May	Jun	Jul	Aug	Sep	Oct	Nov	Budget
Preparation	Kickoff meeting	All		▪								$500
	Create data tables	Lee	Data tables		▭							$4,000
	Prepare data for analysis	Pam, Tony	Prepared data for analysis			▭						$10,000
	Meeting to review data	All	Plan for implementation			▪						$500
Implementation	Find hidden relationships	Pam, Tony	Key facts and trends				▭					$5,000
	Build and optimize model	Pam, Tony	Model with 70% accuracy				▭					$3,000
	Create a report	All	Report					▭				$1,000
	Meet to review	All	Plan for next steps					▪				$500
Deployment	Double blind test	Pam, Lee	Business impact projections						▭			$5,000
	Production rollout	Pam, Lee	Ongoing review							▭		$5,000
	Assess project	All	Project assessment report								▪	$1,000
												$35,500

Figure 2.1. Project timeline

Table 2.1. Project definitions summary

Steps	Details
Define objectives	• Define the business objectives • Define specific and measurable success criteria • Broadly describe the problem • Divide the problem into sub-problems that are unambiguous and that can be solved using the available data • Define the target population • If the available data does not reflect a sample of the target population, generate a plan to acquire additional data
Define deliverables	• Define the deliverables, e.g., a report, new software, business processes, etc. • Understand any accuracy requirements • Define any time-to-compute issues • Define any window-of-opportunity considerations • Detail if and how explanations should be presented • Understand any deployment issues
Define roles and responsibilities	• Project leader • Subject matter expert/business analyst • Data analysis/data mining expert • IT expert • Consumer
Assess current situation	• Define data sources and locations • List assumptions about the data • Understand project constraints (e.g., hardware, software, personnel, etc.) • Assess any legal, privacy or other issues relating to the presentation of the results
Define timetable	• Set aside time for education upfront • Estimate time for the data preparation, implementation, and deployment steps • Set aside time for reviews • Understand risks in the timeline and develop contingency plans
Analyze cost/benefit	• Generate a budget for the project • List the benefits to the business of a successful project • Compare costs and benefits

2.8 FURTHER READING

This chapter has focused on issues relating to large and potentially complex data analysis and data mining projects. There are a number of publications that provide a more detailed treatment of general project management issues including Berkun (2005), Kerzner (2006), and the Project Management Institute's "A Guide to the Project Management Body of Knowledge."

Chapter 3

Preparation

3.1 OVERVIEW

Preparing the data is one of the most time-consuming parts of any data analysis/data mining project. This chapter outlines concepts and steps necessary to prepare a data set prior to any data analysis or data mining exercise. How the data is collected and prepared is critical to the confidence with which decisions can be made. The data needs to be pulled together into a table. This may involve integration of the data from multiple sources. Once the data is in a tabular format it should be fully characterized. The data should also be cleaned by resolving any ambiguities, errors, and removing redundant and problematic data. Certain columns of data can be removed if it is obvious that they would not be useful in any analysis. For a number of reasons, new columns of data may need to be calculated. Finally, the table should be divided, where appropriate, into subsets that either simplify the analysis or allow specific questions to be answered more easily.

Details concerning the steps taken to prepare the data for analysis should be recorded. This not only provides documentation of the activities performed so far, but also provides a methodology to apply to a similar data set in the future. In addition, the steps will be important when validating the results since these records will show any assumptions made about the data.

The following chapter outlines the process of preparing data for analysis. It includes information on the sources of data along with methods for characterizing, cleaning, transforming, and partitioning the data.

3.2 DATA SOURCES

The quality of the data is the single most important factor to influence the quality of the results from any analysis. The data should be reliable and represent the defined target population. Data is often collected to answer specific questions using the following types of studies:

Making Sense of Data: A Practical Guide to Exploratory Data Analysis and Data Mining,
By Glenn J. Myatt

- **Surveys or polls:** A survey or poll can be useful for gathering data to answer specific questions. An interview using a set of predefined questions is usually conducted either over the phone, in person or over the Internet. They are often used to elicit information on people's opinions, preferences and behavior. For example, a poll may be used to understand how a population of eligible voters will cast their vote in an upcoming election. The specific questions to be answered along with the target population should be clearly defined prior to any survey. Any bias in the survey should be eliminated. To achieve this, a true random sample of the target population should be taken. Bias can be introduced in situations where only those responding to the questionnaire are included in the survey since this group may not represent an unbiased random sample. The questionnaire should contain no leading questions, that is, questions that favor a particular response. It is also important that no bias relating to the time the survey was conducted, is introduced. The sample of the population used in the survey should be large enough to answer the questions with confidence. This will be described in more detail within the chapter on statistics.

- **Experiments:** Experiments measure and collect data to answer a specific question in a highly controlled manner. The data collected should be reliably measured, that is, repeating the measurement should not result in different values. Experiments attempt to understand cause and affect phenomena by controlling other factors that may be important. For example, when studying the effects of a new drug, a double blind study is usually used. The sample of patients selected to take part in the study is divided into two groups. The new drug will be delivered to one group, whereas a placebo (a sugar pill) is given to the other group. Neither the patient nor the doctor administering the treatment knows which group the patient is in to avoid any bias in the study on the part of the patient or the doctor.

- **Observational and other studies:** In certain situations it is impossible on either logistical or ethical grounds to conduct a controlled experiment. In these situations, a large number of observations are measured and care taken when interpreting the results.

As part of the daily operations of an organization, data is collected for a variety of reasons. Examples include

- **Operational databases:** These databases contain ongoing business transactions. They are accessed constantly and updated regularly. Examples include supply chain management systems, customer relationship management (CRM) databases and manufacturing production databases.

- **Data warehouses:** A data warehouse is a copy of data gathered from other sources within an organization that has been cleaned, normalized, and optimized for making decisions. It is not updated as frequently as operational databases.

- **Historical databases:** Databases are often used to house historical polls, surveys and experiments.
- **Purchased data:** In many cases data from in-house sources may not be sufficient to answer the questions now being asked of it. One approach is to combine this internal data with data from other sources.

Pulling data from multiple sources is a common situation in many data mining projects. Often the data has been collected for a totally different purpose than the objective of the data mining exercise it is currently being used for. This introduces a number of problems for the data mining team. The data should be carefully prepared prior to any analysis to ensure that it is in a form to answer the questions now being asked. The data should be prepared to mirror as closely as possible the target population about which the questions will be asked. Since multiple sources of data may now have been used, care must be taken bringing these sources together since errors are often introduced at this time. Retaining information on the source of the data can also be useful in interpreting the results.

3.3   DATA UNDERSTANDING

3.3.1   Data Tables

All disciplines collect data about things or objects. Medical researchers collect data on patients, the automotive industry collects data on cars, retail companies collect data on transactions. Patients, cars and transactions are all objects. In a data set there may be many *observations* for a particular object. For example, a data set about cars may contain many observations on different cars. These observations can be described in a number of ways. For example, a car can be described by listing the vehicle identification number (VIN), the manufacturer's name, the weight, the number of cylinders, and the fuel efficiency. Each of these features describing a car is a *variable*. Each observation has a specific value for each variable. For example, a car may have:

VIN = IM8GD9A_KP042788

Manufacturer = Ford

Weight = 2984 pounds

Number of cylinders = 6

Fuel efficiency = 20 miles per gallon

Data sets used for data analysis/data mining are almost always described in tables. An example of a table describing cars is shown in Table 3.1. Each row of the table describes an observation (a specific car). Each column describes a variable (a specific attribute of a car). In this example, there are two observations and these observations are described using five variables: **VIN**, **Manufacturer**, **Weight**,

Table 3.1.   Example of a table describing cars

VIN	Manufacturer	Weight	Number of cylinders	Fuel efficiency
IM8GD9A_KP042788	Ford	2984	6	20
IC4GE9A_DQ1572481	Toyota	1795	4	34

Number of cylinders and **Fuel efficiency**. Variables will be highlighted throughout the book in bold.

A generalized version of the table is shown in Table 3.2. This table describes a series of observations (from O_1 to O_n). Each observation is described using a series of variables ($\mathbf{X_1}$ to $\mathbf{X_k}$). A value is provided for each variable of each observation. For example, the value of the first observation for the first variable is x_{11}.

Getting to the data tables in order to analyze the data may require generating the data from scratch, downloading data from a measuring device or querying a database (as well as joining tables together or pivoting tables), or running a computer software program to generate further variables for analysis. It may involve merging the data from multiple sources. This step is often not trivial. There are many resources describing how to do this, and some are described in the further reading section of this chapter.

Prior to performing any data analysis or data mining, it is essential to thoroughly understand the data table, particularly the variables. Many data analysis techniques have restrictions on the types of variables that they are able to process. As a result, these techniques may be eliminated from consideration or the data must be transformed into an appropriate form for analysis. In addition, certain characteristics of the variables have implications in terms of how the results of the analysis will be interpreted. The following four sections detail a number of ways of characterizing variables.

3.3.2   Continuous and Discrete Variables

A useful initial categorization is to define each variable in terms of the type of values that the variable can take. For example, does the variable contain a fixed number of

Table 3.2.   General format for a table of observations

Observations	Variables				
	x_1	x_2	x_3	...	x_k
O_1	x_{11}	x_{12}	x_{13}	...	x_{1k}
O_2	x_{21}	x_{22}	x_{23}	...	x_{2k}
O_3	x_{31}	x_{32}	x_{33}	...	x_{3k}
...	...	...	...	...	...
O_n	x_{n1}	x_{n2}	x_{n3}	...	x_{nk}

distinct values or could it take any numeric value? The following is a list of descriptive terms for categorizing variables:

- **Constant:** A variable where every data value is the same. In many definitions, a variable must have at least two different values; however, it is a useful categorization for our purposes. For example, a variable **Calibration** may indicate the value a machine was set to in order to generate a particular measurement and this value may be the same for all observations.

- **Dichotomous:** A variable where there are only two values, for example, **Gender** whose values can be male or female. A special case is a *binary* variable whose values are 0 and 1. For example, a variable **Purchase** may indicate whether a customer bought a particular product and the convention that was used to represent the two cases is 0 (did not buy) and 1 (did buy).

- **Discrete:** A variable that can only take a certain number of values (either text or numbers). For example, the variable **Color** where values could be black, blue, red, yellow, and so on, or the variable **Score** where the variable can only take values 1, 2, 3, 4, or 5.

- **Continuous:** A variable where an infinite number of numeric values are possible within a specific range. An example of a continuous value is **temperature** where between the minimum and maximum temperature, the variable could take any value.

It can be useful to describe a variable with additional information. For example, is the variable a count or fraction, a time or date, a financial term, a value derived from a mathematical operation on other variables, and so on? The units are also useful information to capture in order to present the result. When two tables are merged, units should also be aligned or appropriate transformations applied to ensure all values have the same unit.

3.3.3 Scales of Measurement

The variable's scale indicates the accuracy at which the data has been measured. This classification has implications as to the type of analysis that can be performed on the variable. The following terms categorize scales:

- **Nominal:** Scale describing a variable with a limited number of different values. This scale is made up of the list of possible values that the variable may take. It is not possible to determine whether one value is larger than another. For example, a variable **Industry** would be nominal where it takes values such as financial, engineering, retail, and so on. The order of these values has no meaning.

- **Ordinal:** This scale describes a variable whose values are ordered; however, the difference between the values does not describe the magnitude of the actual difference. For example, a scale where the only values are low, medium, and high tells us that high is larger than medium, and medium is

Table 3.3. Scales of measurement summary

	Meaningful order	Meaningful difference	Natural zero
Nominal	No	No	No
Ordinal	Yes	No	No
Interval	Yes	Yes	No
Ratio	Yes	Yes	Yes

larger than low. However, it is impossible to determine the magnitude of the difference between the three values.

- **Interval:** Scales that describe values where the interval between the values has meaning. For example, when looking at three data points measured on the Fahrenheit scale, 5 °F, 10 °F, 15 °F, the differences between the values from 5 to 10 and from 10 to 15 are both 5 and a difference of 5 °F in both cases has the same meaning. Since the Fahrenheit scale does not have a lowest value at zero, a doubling of a value does not imply a doubling of the actual measurement. For example, 10 °F is not twice as hot as 5 °F. Interval scales do not have a natural zero.

- **Ratio:** Scales that describe variables where the same difference between values has the same meaning (as in interval) but where a double, tripling, etc. of the values implies a double, tripling, etc. of the measurement. An example of a ratio scale is a bank account balance whose possible values are $5, $10, and $15. The difference between each pair is $5 and $10 is twice as much as $5. Since ratios of values are possible, they are defined as having a natural zero.

Table 3.3 provides a summary of the different types of scales.

3.3.4 Roles in Analysis

It is also useful to think about how the variables will be used in any subsequent analysis. Example roles in data analysis and data mining include

- **Labels:** Variables that describe individual observations in the data.

- **Descriptors:** These variables are almost always collected to describe an observation. Since they are often present, these variables are used as the input or *descriptors* to be used in both creating a predictive model and generating predictions from these models. They are also described as predictors or X variables.

- **Response:** These variables (usually one variable) are predicted from a predictive model (using the descriptor variables as input). These variables will be used to guide the creation of the predictive model. They will also be predicted, based on the input descriptor variables that are presented to the model. They are also referred to as Y variables.

The car example previously described had the following variables: **vehicle identification number (VIN)**, **Manufacturer**, **Weight**, **Number of cylinders**, and **Fuel efficiency**. One way of using this data is to build a model to predict **Fuel efficiency**. The **VIN** variable describes the individual observations and is assigned as a label. The variables **Manufacturer**, **Weight**, and **Number of cylinders** will be used to create a model to predict **Fuel efficiency**. Once a model is created, the variables **Manufacturer**, **Weight**, and **Number of cylinders** will be used as inputs to the model and the model will predict **Fuel efficiency**. The variables **Manufacturer**, **Weight**, and **Number of cylinders** are descriptors, and the variable **Fuel efficiency** is the response variable.

3.3.5 Frequency Distribution

For variables with an ordered scale (ordinal, interval, or ratio), it is useful to look at the frequency distribution. The frequency distribution is based on counts of values or ranges of values (in the case of interval or ratio scales). The following histogram shows a frequency distribution for a variable **X**. The variable has been classified into a series of ranges from -6 to -5, -5 to -4, -4 to -3, and so on, and the graph in Figure 3.1 shows the number of observations for each range. It indicates that the majority of the observations are grouped in the middle of the distribution between -2 and $+1$, and there are relatively fewer observations at the extreme values. The frequency distribution has an approximate bell-shaped curve as shown in Figure 3.2. A symmetrical bell-shaped distribution is described as a *normal* (or Gaussian) distribution. It is very common for variables to have a normal distribution. In addition, many data analysis techniques assume an approximate normal distribution. These techniques are referred to as *parametric* procedures (*nonparametric* procedures do not require a normal distribution).

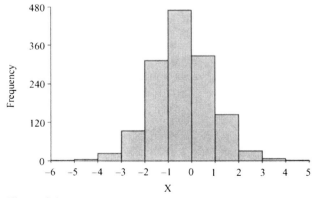

Figure 3.1. Frequency distribution for variable **X**

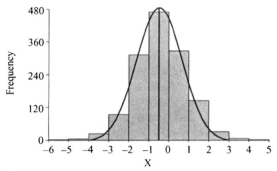

Figure 3.2.   Frequency distribution for variable **X** with the normal distribution superimposed

3.4   DATA PREPARATION

3.4.1   Overview

Having performed a preliminary data characterization, it is now time to analyze further and transform the data set prior to starting any analysis. The data must be cleaned and translated into a form suitable for data analysis and data mining. This process will enable us to become familiar with the data and this familiarity will be beneficial to the analysis performed in step 3 (the implementation of the analysis). The following sections review some of the criteria and analysis that can be performed.

3.4.2   Cleaning the Data

Since the data available for analysis may not have been originally collected with this project's goal in mind, it is important to spend time cleaning the data. It is also beneficial to understand the accuracy with which the data was collected as well as correcting any errors.

For variables measured on a nominal or ordinal scale (where there are a fixed number of possible values), it is useful to inspect all possible values to uncover mistakes and/or inconsistencies. Any assumptions made concerning possible values that the variable can take should be tested. For example, a variable **Company** may include a number of different spellings for the same company such as:

General Electric Company

General Elec. Co

GE

Gen. Electric Company

General electric company

G.E. Company

These different terms, where they refer to the same company, should be consolidated into one for analysis. In addition, subject matter expertise may be needed in cleaning these variables. For example, a company name may include one of the divisions of the General Electric Company and for the purpose of this specific project it should be included as the "General Electric Company."

It can be more challenging to clean variables measured on an interval or ratio scale since they can take any possible value within a range. However, it is useful to consider outliers in the data. Outliers are a single or a small number of data values that are not similar to the rest of the data set. There are many reasons for outliers. An outlier may be an error in the measurement. A series of outlier data points could be a result of measurements made using a different calibration. An outlier may also be a genuine data point. Histograms, scatterplots, box plots and z-scores can be useful in identifying outliers and are discussed in more detail within the next two chapters. The histogram in Figure 3.3 displays a variable **Height** where one value is eight times higher than the average of all data points.

There are additional methods such as clustering and regression that could also be used to identify outliers. These methods are discussed later in the book. Diagnosing an outlier will require subject matter expertise to determine whether it is an error (and should be removed) or a genuine observation. If the value or values are correct, then the variable may need some mathematical transformation to be applied for use with data analysis and data mining techniques. This will be discussed later in the chapter.

Another common problem with continuous variables is where they include nonnumeric terms. Any term described using text may appear in the variable, such as "above 50" or "out of range." Any numeric analysis would not be able to interpret a value that is not an explicit number, and hence, these terms should be converted to a number, based on subject matter expertise, or should be removed.

In many situations, an individual observation may have data missing for a particular variable. Where there is a specific meaning for a missing data value, the value may be replaced on the basis of the knowledge of how the data was collected.

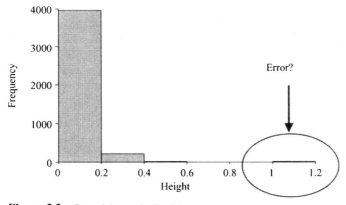

Figure 3.3. Potential error in the data

Alternatively, the observation should be removed from the table. There are methods that attempt to estimate a value for missing data; however, these methods should be used with care. Literature describing these methods has been listed in the further reading section of this chapter.

A particular variable may have been measured over different units. For example, a variable **Weight** may have been measured using both pounds and kilograms for different observations and should be standardized to a single scale. Another example would be where a variable **Price** is shown in different currencies and should be standardized to one for the purposes of analysis. In situations where data has been collected over time, there may be changes related to the passing of time that is not relevant for the analysis. For example, when looking at a variable **Cost of production** where the data has been collected over many years, the rise in costs attributable to inflation may need to be factored out for this specific analysis.

By combining data from multiple sources, an observation may have been recorded more than once and any duplicate entries should be removed.

3.4.3   Removing Variables

On the basis of an initial categorization of the variables, it may be possible to remove variables from consideration at this point. For example, constants and variables with too many missing data points should be considered for removal. Further analysis of the correlations between multiple variables may identify variables that provide no additional information to the analysis and hence could be removed. This type of analysis is described in the chapter on statistics.

3.4.4   Data Transformations

Overview

It is important to consider applying certain mathematical transformations to the data since many data analysis/data mining programs will have difficulty making sense of the data in its raw form. Some common transformations that should be considered include normalization, value mapping, discretization, and aggregation. When a new variable is generated, the transformation procedure used should be retained. The inverse transformation should then be applied to the variable prior to presenting any analysis results that include this variable. The following section describes a series of data transformations to apply to data sets prior to analysis.

Normalization

Normalization is a process where numeric columns are transformed using a mathematical function to a new range. It is important for two reasons. First, any analysis of the data should treat all variables equally so that one column does not have more influence over another because the ranges are different. For example, when analyzing customer credit card data, the **Credit limit** value is not given more weightage in the analysis than the **Customer's age**. Second, certain data analysis and data mining methods require the data to be normalized prior to analysis, such as neural networks or k-nearest neighbors, described in Sections 7.3 and 7.5. The following outlines some common normalization methods:

- **Min-max:** Transforms the variable to a new range, such as from 0 to 1. The following formula is used:

$$Value' = \frac{Value - OriginalMin}{OriginalMax - OriginalMin}(NewMax - NewMin) + NewMin$$

where $Value'$ is the new normalized value, $Value$ is the original variable value, $OriginalMin$ is the minimum possible value of the original variable, $OriginalMax$ is the maximum original possible value, $NewMin$ is the minimum value for the normalized range, and $NewMax$ is the maximum value for the normalized range. This is a useful formula that is widely used. The minimum and maximum values for the original variable are needed. If the original data does not contain the full range, either a best guess at the range is needed or the formula should be restricted for future use to the range specified.

- **z-score:** It normalizes the values around the mean (or average) of the set, with differences from the mean being recorded as standardized units on the basis of the frequency distribution of the variable. The following formula is used:

$$Value' = \frac{Value - \bar{x}}{s}$$

where $\bar{x}$ is the mean or average value for the variable and s is the standard deviation for the variable. Calculations and descriptions for mean and standard deviation calculations are provided in the chapter on statistics.

- **Decimal scaling:** This transformation moves the decimal to ensure the range is between 1 and −1. The following formula is used:

$$Value' = \frac{Value}{10^n}$$

Where n is the number of digits of the maximum absolute value. For example, if the largest number is 9948 then n would be 4. 9948 would normalize to $9948/10^4$, 9948/10,000, or 0.9948.

The normalization process is illustrated using the data in Table 3.4. To calculate the normalized values using the min-max equation, first the minimum and maximum values should be identified: $OriginalMin = 7$ and $OriginalMax = 53$.

Table 3.4.   Single column to be normalized

Variable
33
21
7
53
29
42
12
19
22
36

The new normalized values will be between 0 and 1: $NewMin = 0$ and $NewMax = 1$. To calculate the new normalized value ($value'$) using the formula for the value 33:

$$Value' = \frac{Value - OriginalMin}{OriginalMax - OriginalMin} \ (NewMax - NewMin) + NewMin$$

$$Value' = \frac{33 - 7}{53 - 7} \ (1 - 0) + 0$$

$$Value' = 0.565$$

Table 3.5 shows the calculated normalized values for all data points.

A variable may not conform to a normal distribution. Certain data analysis methods require the data to follow a normal distribution. Methods for visualizing and describing a normal frequency distribution are described in the following two chapters. To transform the data into a more appropriate normal distribution, it may be necessary to take the log (or negative log), exponential or perform a Box-Cox

Table 3.5.   Variable normalized to the range 0–1

Variable	Normalized (0 to 1)
33	0.565
21	0.304
7	0
53	1
29	0.478
42	0.761
12	0.109
19	0.261
22	0.326
36	0.630

Table 3.6.   Example of transformation to generate a normal distribution

Value	Exp (Value)
5.192957	180
5.799093	330
6.063785	430
6.068426	432

transformation. The formula for a Box-Cox transformation is:

$$Value' = \frac{Value^{\lambda} - 1}{\lambda}$$

where λ is a value greater than 1.

In Table 3.6 the original variable is transformed using an exponential function and the distribution is now more normal (see Figure 3.4). The table shows a sample of the original (**Value**) and the newly calculated column: **Exp (Value)**.

Value Mapping

To use variables that have been assigned as ordinal and described using text values within certain numerical analysis methods, it will be necessary to convert the variable's values into numbers. For example, a variable with low, medium, and high values may have low values replaced by 0, medium values replaced by 1, and high values replaced by 2. However, this conversion should be approached with care and with as much subject matter expertise as possible to assign the appropriate score to each value.

Another approach to handling nominal data is to convert each value into a separate column with values 1 (indicating the presence of the category) and 0 (indicating the absence of the category). These new variables are often referred to as

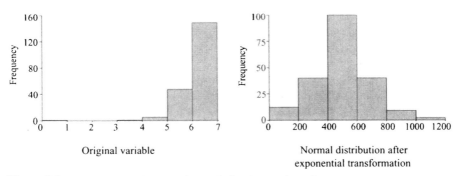

Original variable

Normal distribution after exponential transformation

Figure 3.4.   Frequency distribution before and after the transformation

Table 3.7.   Mapping nominal data onto a series of dummy variables

Original column	New variables (value-mapping)				
Color	Color = Red	Color = Green	Color = Blue	Color = Orange	Color = Yellow
red	1	0	0	0	0
green	0	1	0	0	0
blue	0	0	1	0	0
red	1	0	0	0	0
blue	0	0	1	0	0
orange	0	0	0	1	0
yellow	0	0	0	0	1
red	1	0	0	0	0

dummy variables. For example, the variable **Color** has now been divided into 5 separate columns, one for each value, as shown in Table 3.7.

Discretization

By converting continuous data into discrete values, it would appear that we are loosing information. However, this conversion is desirable in a number of situations. Firstly, where a value is defined on an interval or ratio scale but when knowledge about how the data was collected suggests the accuracy of the data does not warrant these scales, a variable may be a candidate for *discretization*. This is often referred to as *data smoothing*. It may be more desirable to convert the data into more broadly defined categories that reflect the true variation in the data. Secondly, certain techniques can only process categorical data and hence converting continuous data into discrete values makes the variable accessible to these methods. For example, a continuous variable **Credit score** may be divided into four categories: poor, average, good and excellent.

This type of conversion or *binning* can be illustrated with an example. A variable **Weight** that has a range from 0 to 350 lbs may be divided into five categories: less than 100 lb, 100–150 lb, 150–200 lb, 200–250 lb and above 250 lb. All values for the variable **Weight** must now be assigned to a category and assigned an appropriate value such as the mean of the assigned category. It is often useful to use the frequency distribution to understand appropriate binning cut-offs.

Discretization can also be applied to nominal variables. This is often useful in situations where there is a large number of values for a given nominal variable. If the data set were to be summarized using each of the values, the number of observations for each value may be too small to meaningfully reach any conclusions. However, a new column could be generated that generalizes the values using a mapping of terms. For example, a data set concerning customer transactions may contain a variable **Company** that details the individual customer's company. There may only be a handful of observations for each company. However, this variable could be mapped onto a new variable, **Industries**. The mapping of specific companies onto generalized industries must be defined using a concept mapping (i.e., which company maps onto which industry). Now, when the data set is summarized using the values for the **Industries** variable, meaningful trends may be seen.

Aggregation

The variable that you are trying to use may not be present in the data set, but it may be derived from other variables present. Any mathematical operation, such as average or sum, could be applied to one or more variables in order to create an additional variable. For example, a project may be trying to understand issues around a particular car's fuel efficiency (**Fuel Efficiency**) using a data set of different journeys where the fuel level at the start (**Fuel Start**) and the end (**Fuel End**) of a trip is measured along with the distance covered (**Distance**). An additional column may be calculated using the following formula:

$$\textbf{Fuel Efficiency} = (\textbf{Fuel End} - \textbf{Fuel Start})/\textbf{Distance}$$

3.4.5 Segmentation

Generally, larger data sets take more computational time to analyze. Segmenting (creating subsets) the data can speed up any analysis. One approach is to take a random subset. This approach is effective where the data set closely matches the target population. Another approach is to use the problem definition to guide how the subset is constructed. For example, a problem may have been defined as: analyze an insurance dataset of 1 million records to identify factors leading to fraudulent claims. The data set may only contain 20,000 fraudulent claims. Since it will be essential to compare fraudulent and nonfraudulent claims in the analysis, it will be important to create a data set of examples of both. The 20,000 fraudulent claims could be combined with a random sample of 20,000 nonfraudulent claims. This process will result in a smaller subset for analysis.

A data set may have been built up over time and collected to answer a series of questions. Now, this data set may be used for a different purpose. It may be necessary to select a diverse set of observations that more closely matches the new target population. For example, a car safety organization has been measuring the safety of individual cars on the basis of specific requests from the government. Over time, the government may have requested car safety studies for certain types of vehicles. Now, if this historical data set is to be used to answer questions on the safety of all cars, this data set does not reflect the new target population. However, a subset of the car studies could be selected to represent the more general questions now being asked of the data. The chapter on grouping will discuss how to create diverse data sets when the data does not represent the target population.

When building predictive models from a data set, it is important to keep the models as simple as possible. Breaking the data set down into subsets based on

Table 3.8.   Summary of the steps when preparing data

Steps	Details
1. Create data table	• Query databases to access data • Integrate multiple data sets and format as a data table
2. Characterize variables	Characterize the variables based on: • Continuous/discrete • Scales of measurement • Roles in analysis • Frequency distribution
3. Clean data	Clean the data: • Consolidate observations by merging appropriate terms • Identify potential errors (outliers, non-numeric characters, etc.) • Appropriately set nonnumeric values (or remove) • Ensure measurements are taken over the same scale • Remove duplicate observations
4. Remove variables	Remove variables that will not contribute to any analysis (e.g., constants or variables with too few values)
5. Transform variables	Transform the variable, if necessary, retaining how the variable was transformed using the following operations: • Normalize • Value mapping • Discretization • Aggregation
6. Segment table	Create subsets of the data to: • Facilitate more rapid analysis • Simplify the data set to create simpler models • Answer specific questions

your knowledge of the data may allow you to create multiple but simpler models. For example, a project to model factors that contribute to the price of real estate may use a data set of nationwide house prices and associated factors. However, your knowledge of the real estate market suggests that factors contributing to house prices are contingent on the area. Factors that contribute to house prices in coastal locations are different from factors that contribute to house prices in the mountains. It may make sense, in this situation, to divide the data into smaller sets based on location and to model these locales separately. When doing this type of subsetting, it is important to note the criteria you are using to subset the data. These criteria will be needed when data to be predicted is presented for modeling by assigning the data to one or more models. In situations where multiple predictions are generated for the same unknown observation, a method for consolidating these predictions will be required. This topic will be discussed further in the chapter on prediction.

3.5 SUMMARY

Table 3.8 details the steps and issues from this stage of the project. The steps performed on the data should be documented. The deliverables from this stage in the project are a prepared data set for analysis as well as a thorough understanding of the data.

3.6 EXERCISES

A set of 10 hypothetical patient records from a large database is presented in Table 3.9. Patients with a diabetes value of 1 have type II diabetes and patients with a diabetes value of 0 do not have type II diabetes. It is anticipated that this data set will be used to predict diabetes based on measurements of age, systolic blood pressure, diastolic blood pressure, and weight.

1. For the following variables from Table 3.9, assign them to one of the following categories: constant, dichotomous, binary, discrete, and continuous.
   a. **Name**
   b. **Age**
   c. **Gender**
   d. **Blood group**
   e. **Weight** (kg)
   f. **Height** (m)
   g. **Systolic blood pressure**
   h. **Diastolic blood pressure**
   i. **Temperature**
   j. **Diabetes**

2. For each of the following variables, assign them to one of the following scales: nominal, ordinal, interval, ratio.
   a. **Name**
   b. **Age**
   c **Gender**
   d. **Blood group**
   e. **Weight** (kg)
   f. **Height** (m)
   g. **Systolic blood pressure**
   h. **Diastolic blood pressure**
   i. **Temperature**
   j. **Diabetes**

3. On the basis of the anticipated use of the data to build a predictive model, identify:
   a. A label for the observations
   b. The descriptor variables
   c. The response variable

4. Create a new column by normalizing the **Weight** (kg) variable into the range 0 to 1 using the min-max normalization.

5. Create a new column by binning the **Weight** variable into three categories: low (less than 60 kg), medium (60–100 kg), and high (greater than 100 kg).

Table 3.9. Table of patient records

Name	Age	Gender	Blood group	Weight (kg)	Height (m)	Systolic blood pressure	Diastolic blood pressure	Temperature (°F)	Diabetes
P. Lee	35	Female	A Rh$^+$	50	1.52	68	112	98.7	0
R. Jones	52	Male	O Rh$^-$	115	1.77	110	154	98.5	1
J. Smith	45	Male	O Rh$^+$	96	1.83	88	136	98.8	0
A. Patel	70	Female	O Rh$^-$	41	1.55	76	125	98.6	0
M. Owen	24	Male	A Rh$^-$	79	1.82	65	105	98.7	0
S. Green	43	Male	O Rh$^-$	109	1.89	114	159	98.9	1
N. Cook	68	Male	A Rh$^+$	73	1.76	108	136	99.0	0
W. Hands	77	Female	O Rh$^-$	104	1.71	107	145	98.3	1
P. Rice	45	Female	O Rh$^+$	64	1.74	101	132	98.6	0
F. Marsh	28	Male	O Rh$^+$	136	1.78	121	165	98.7	1

6. Create an aggregated column, body mass index (**BMI**), which is defined by the formula:

$$\mathbf{BMI} = \frac{\mathbf{Weight}\ (\mathrm{kg})}{\mathbf{Height}\ (\mathrm{m})^2}$$

7. Segment the data into data sets based on values for the variable **Gender**.

3.7 FURTHER READING

This chapter has reviewed some of the sources of data used in exploratory data analysis and data mining. The following books provide more information on surveys and polls: Fowler (2002), Rea (2005), and Alreck (2004). There are many additional resources describing experimental design including Montgomery (2005), Cochran (1957), Barrentine (1999), and Antony (2003). Operational databases and data warehouses are summarized in the following books: Oppel (2004) and Kimball (2002). Oppel (2004) also summarizes access and manipulation of information in databases. Principal component analysis provides the opportunity to reduce the number of variables into a smaller set of principal components and is often used as a data reduction method. It is outlined in Jolliffe (2002) and Jackson (2003). For additional data preparation approaches including the handling of missing data see Pearson (2005), Pyle (1999), and Dasu (2003).

Chapter 4

Tables and Graphs

4.1 INTRODUCTION

The following chapter describes a series of techniques for summarizing data using tables and graphs. Tables can be used to present both detailed and summary level information about a data set. Graphs visually communicate information about variables in data sets and the relationship between them. The following chapter describes a series of tables and graphs useful for exploratory data analysis and data mining.

4.2 TABLES

4.2.1 Data Tables

The most common way of looking at data is through a table, where the raw data is displayed in familiar rows of observations and columns of variables. It is essential for reviewing the raw data; however, the table can be overwhelming with more than a handful of observations or variables. Sorting the table based on one or more variables is useful for organizing the data. It is virtually impossible to identify any trends or relationships looking at the raw data alone. An example of a table describing different cars is shown in Table 4.1.

4.2.2 Contingency Tables

Contingency tables (also referred to as two-way cross-classification tables) provide insight into the relationship between two variables. The variables must be categorical (dichotomous or discrete), or transformed to a categorical variable. A variable is often dichotomous; however, a contingency table can represent variables with more than two values. Table 4.2 describes the format for a contingency table where two variables are compared: **Variable x** and **Variable y**.

Making Sense of Data: A Practical Guide to Exploratory Data Analysis and Data Mining,
By Glenn J. Myatt
Copyright © 2007 John Wiley & Sons, Inc.

Table 4.1.   Table of car records

Names	Cylinders	Displace-ment	Horse-power	Weight	Acce-leration	Model Year	Origin	MPG
Chevrolet Chevelle Malibu	8	307	130	3504	12	1970	1	18
Buick Skylark 320	8	350	165	3693	11.5	1970	1	15
Plymouth Satellite	8	318	150	3436	11	1970	1	18
Amc Rebel SST	8	304	150	3433	12	1970	1	16
Ford Torino	8	302	140	3449	10.5	1970	1	17
Ford Galaxie 500	8	429	198	4341	10	1970	1	15
Chevrolet Impala	8	454	220	4354	9	1970	1	14
Plymouth Fury III	8	440	215	4312	8.5	1970	1	14
Pontiac Catalina	8	455	225	4425	10	1970	1	14
Amc Ambassador Dpl	8	390	190	3850	8.5	1970	1	15

- $Count_{+1}$: the number of observations where **Variable x** has "Value 1", irrespective of the value of **Variable y**.
- $Count_{+2}$: the number of observations where **Variable x** has "Value 2", irrespective of the value of **Variable y**.
- $Count_{1+}$: the number of observations where **Variable y** has "Value 1", irrespective of the value of **Variable x**.
- $Count_{2+}$: the number of observations where **Variable y** has "Value 2", irrespective of the value of **Variable x**.

The total number of observations in the data set is shown as *Total count*. The number of observations where the value of **Variable x** equals "Value 1" and the value of **Variable y** equals "Value 1" is shown in the cell $Count_{11}$. $Count_{21}$, $Count_{12}$, and $Count_{22}$ show counts for the overlaps between all other values. The counts can also be annotated and/or replaced with percentages.

In Table 4.3, the data set is summarized using two variables: **sex** and **age**. The variable **sex** is dichotomous and the two values (male and female) are shown as a header on the *x*-axis. The other selected variable is **age** and has been broken down into nine categories: 10–20, 20–30, 30–40, and so on. For each value of each variable a total is displayed. For example, there are 21,790 observations where **sex** is equal to

Table 4.2.   Contingency table format

		Variable x		Totals
		Value 1	Value 2	
Variable y	Value 1	$Count_{11}$	$Count_{21}$	$Count_{1+}$
	Value 2	$Count_{12}$	$Count_{22}$	$Count_{2+}$
		$Count_{+1}$	$Count_{+2}$	*Total count*

Table 4.3. Contingency table summarizing the number of males and females within age ranges

	Sex = Male	Sex = Female	Totals
Age (10.0 to 20.0)	847	810	1,657
Age (20.0 to 30.0)	4,878	3,176	8,054
Age (30.0 to 40.0)	6,037	2,576	8,613
Age (40.0 to 50.0)	5,014	2,161	7,175
Age (50.0 to 60.0)	3,191	1,227	4,418
Age (60.0 to 70.0)	1,403	612	2,015
Age (70.0 to 80.0)	337	171	508
Age (80.0 to 90.0)	54	24	78
Age (90.0 to 100.0)	29	14	43
Totals	21,790	10,771	32,561

male and there are 1657 observations where age is between 10 and 20. The total number of observations summarized in the table is shown in the bottom right hand corner (32,561). The cells in the center of the table show the number of observations with different combinations of values. For example, there are 847 males between the ages of 10 and 20 years.

Where one of the variables is a response, it is common to place this variable on the y-axis. In Table 4.4, 392 observations about cars are summarized according to two variables: number of cylinders (**Cylinders**) and miles per gallon (**MPG**). This table describes the relationship between the number of cylinders in a car and the car's fuel efficiency. This relationship can be seen by looking at the relative distribution of observations throughout the grid. In the column where **Cylinders** is 4, the majority of the data lies between 20 and 40 **MPG**. Whereas the column where **Cylinders** is 8, the majority of observations lies between 10 and 20 **MPG**, indicating that 8-cylinder vehicles appear to be less fuel efficient than 4-cylinder vehicles.

Table 4.4. Contingency table summarizing counts of cars based on the number of cylinders and ranges of fuel efficiency (MPG)

	Cylinders = 3	Cylinders = 4	Cylinders = 5	Cylinders = 6	Cylinders = 8	Totals
MPG (5.0 to 10.0)	0	0	0	0	1	1
MPG (10.0 to 15.0)	0	0	0	0	52	52
MPG (15.0 to 20.0)	2	4	0	47	45	98
MPG (20.0 to 25.0)	2	39	1	29	4	75
MPG (25.0 to 30.0)	0	70	1	4	1	76
MPG (30.0 to 35.0)	0	53	0	2	0	55
MPG (35.0 to 40.0)	0	25	1	1	0	27
MPG (40.0 to 45.0)	0	7	0	0	0	7
MPG (45.0 to 50.0)	0	1	0	0	0	1
Totals	4	199	3	83	103	392

Contingency tables have many uses including understanding the relationship between two categorical variables. In the chapter on statistics, this relationship will be further quantified using the chi-square test. They are also useful for looking at the quality of predictions and this will be discussed further in the chapter on prediction.

4.2.3 Summary Tables

A *summary table* (or aggregate table) is a common way of understanding data. For example, a retail company may generate a summary table to communicate the average sales per product or per store. A single categorical variable (or a continuous variable converted into categories) is used to group the observations. Each row of the table represents a single group. Summary tables will often show a count of the number of observations (or percentage) that have that particular value (or range). Any number of other variables can be shown alongside. Since each row now refers to a set of observations, any other columns of variables must now contain summary information. Descriptive statistics that summarize a set of observations can be used. The calculations for these statistics are described in the next chapter. The following statistics are commonly used:

- **Mean:** The average value.
- **Median:** The value at the mid-point.
- **Sum:** The sum over all observations in the group.
- **Minimum:** The minimum value.
- **Maximum:** The maximum value.
- **Standard deviation:** A standardized measure of the deviation of a variable from the mean.

A common format for a summary table is shown in Table 4.5. The first column is the variable used to group the table (**Variable a**). Each value (either a specific value or a range) is listed in the first column alongside a count (or percentage) of observations belonging to the group. Each row now represents a collection of observations. Other columns present summaries for other variables. **Variable x** and **Variable y** are examples of those additional summarized columns.

Table 4.5. Format for a summary table

Variable a	Count	Variable x summary	Variable y summary	...
a_1	Count (a_1)	Statistic(x) for group a_1	Statistic(y) for group a_1	...
a_2	Count (a_2)	Statistic(x) for group a_2	Statistic(y) for group a_2	...
a_3	Count (a_3)	Statistic(x) for group a_3	Statistic(y) for group a_3	...
...	...	...	...	...
a_n	Count (a_n)	Statistic(x) for group a_n	Statistic(y) for group a_n	...

Table 4.6. Summary table showing average **MPG** for different cylinder vehicles

Cylinders	Count	Mean (MPG)
3.0	4	20.55
4.0	199	29.28
5.0	3	27.37
6.0	83	19.97
8.0	103	14.96

In Table 4.6, the automobile data set is broken down into groups based on the number of **Cylinders** (3, 4, 5, 6 and 8). A count of the number of observations in each group is shown in the next column. The third column is based on another variable, miles per gallon (**MPG**), and the statistic selected is mean. The table succinctly summarizes how the average fuel efficiency of the set of automobiles differs based on the number of cylinders in the car.

Summary tables summarize the contents of a data table without showing all the details. It is possible to identify trends and these tables are easy to understand.

4.3 GRAPHS

4.3.1 Overview

Tables allow us to look at individual observations or summaries, whereas graphs present the data visually replacing numbers with graphical elements. Tables are important when the actual data values are important to show. Graphs enable us to visually identify trends, ranges, frequency distributions, relationships, outliers and make comparisons. There are many ways of visualizing information in the form of a graph. This section will describe some of the common graphs used in exploratory data analysis and data mining: frequency polygrams, histograms, scatterplots, and box plots. In addition, looking at multiple graphs simultaneously and viewing common subsets can offer new insights into the whole data set.

4.3.2 Frequency Polygrams and Histograms

Frequency polygrams plot information according to the number of observations reported for each value (or ranges of values) for a particular variable. An example of a frequency polygram is shown in Figure 4.1. In this example, a variable (**Model Year**) is plotted. The number of observations for each year is counted and plotted. The shape of the plot reveals trends, that is, the number of observations each year fluctuates within a narrow range of around 25–40.

In Figure 4.2, a continuous variable (**Displacement**) is divided into ranges from 50 to 100, from 100 to 150, and so on. The number of values for each range is plotted

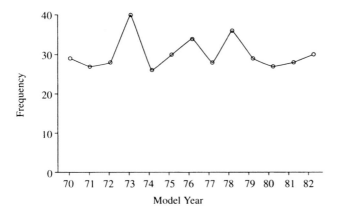

Figure 4.1. Frequency polygram displaying a count for cars per year

and the shape indicates that most of the observations are for low displacement values.

Histograms present very similar information to frequency polygrams, that is, the frequency distribution of a particular variable. The length of the bar is proportional to the size of the group. Variables that are not continuous can be shown as a histogram, as shown in Figure 4.3. This graph shows the dichotomous variable **Diabetes**, which has two values: yes and no. The length of the bars represents the number of observations for the two values. This type of chart for categorical variables is also referred to as a bar chart.

For continuous variables, a histogram can be very useful in displaying the frequency distribution. In Figure 4.4, the continuous variable **Length** is divided into 10 groups and the frequency of the individual group is proportional to the length of the bar.

Histograms provide a clear way of viewing the frequency distribution for a single variable. The central values, the shape, the range of values as well as any outliers can be identified. For example, the histogram in Figure 4.5 illustrates an

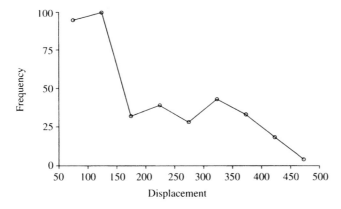

Figure 4.2. Frequency polygram showing counts for ranges of **Displacement**

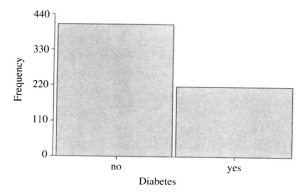

Figure 4.3.   Histogram showing categorical variable **Diabetes**

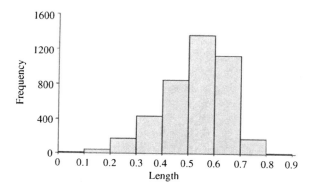

Figure 4.4.   Histogram representing counts for ranges in the variable **Length**

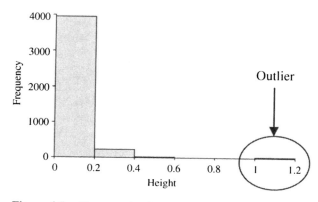

Figure 4.5.   Histogram showing an outlier

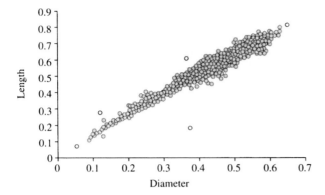

Figure 4.6. Scatterplot showing the relationship between the **Length** and **Diameter** variables

outlier (in the range 1–1.2) that is considerably larger than the majority of other observations. It is also possible to deduce visually if the variable approximates a normal distribution.

4.3.3 Scatterplots

Scatterplots can be used to identify whether any relationship exists between two continuous variables based on the ratio or interval scales. The two variables are plotted on the *x*- and *y*-axes. Each point displayed on the scatterplot is a single observation. The position of the point is determined by the value of the two variables. The scatterplot in Figure 4.6 presents many thousands of observations on a single chart.

Scatterplots allow you to see the type of relationship that may exist between the two variables. For example, the scatterplot in Figure 4.7 shows that the

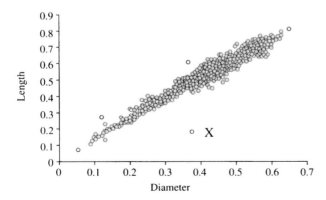

Figure 4.7. Scatterplot showing an outlier (*X*)

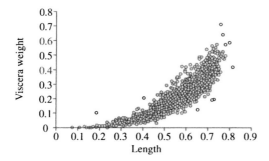

Figure 4.8. Scatterplot showing a nonlinear relationship

relationship between **Length** and **Diameter** is primarily *linear*, that is, as **Length** increases **Diameter** increases proportionally. The graph also shows that there are points (e.g., *X*) that do not follow this linear relationship. These are referred to as outliers based on the dimensions plotted. Where the points follow a straight line or a curve, a simple relationship exists between the two variables. In Figure 4.8, the points follow a curve indicating that there is a *nonlinear* relationship between the two variables, that is, as **Length** increases **Viscera weight** increases, but the rate of increase is not proportional. Scatterplots can also show the lack of any relationship. In Figure 4.9, the points are scattered throughout the whole graph indicating that there is no immediately obvious relationship between **Plasma–Glucose** and **BMI** in this data set. Scatterplots can also indicate where there is a *negative* relationship. For example, it can be seen in Figure 4.10 that as values for **Horsepower** increase, values for **MPG** decrease.

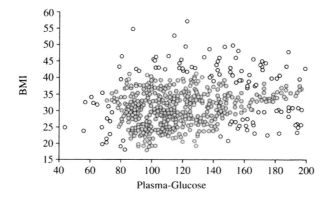

Figure 4.9. Scatterplot showing no discernable relationship

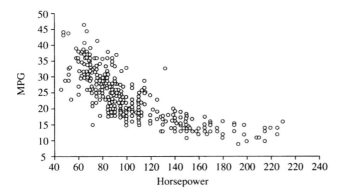

Figure 4.10.   Scatterplot showing a negative relationship

4.3.4 Box Plots

Box plots (also called box-and-whisker plots) provide a succinct summary of the overall distribution for a variable. Five points are displayed: the lower extreme value, the lower quartile, the median, the upper quartile, the upper extreme and the mean, as shown in Figure 4.11. The values on the box plot are defined as follows:

- **Lower extreme:** The lowest value for the variable.
- **Lower quartile:** The point below which 25% of all observations fall.
- **Median:** The point below which 50% of all observations fall.
- **Upper quartile:** The point below which 75% of all observations fall.
- **Upper extreme:** The highest value for the variable.
- **Mean:** The average value for the variable.

Figure 4.12 provides an example of a box plot for one variable (**MPG**). The plot visually displays the lower (around 9) and upper (around 47) bounds of the variable.

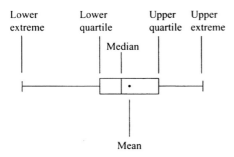

Figure 4.11.   Box plot format

5   10   15   20   25   30   35   40   45   50

Figure 4.12.   Example of box plot for the variable **MPG**

Fifty percent of observations begin at the lower quartile (around 17) and end at the upper quartile (around 29). The median and the mean values are close, with the mean slightly higher (around 23.5) than the median (around 23). Figure 4.13 shows a box plot and a histogram side-by-side to illustrate how the distribution of a variable is summarized using the box plot.

In certain version of the box plot, outliers are not included in the plot. These extreme values are replaced with the highest and lowest values not considered as an outlier. Instead these outliers are explicitly drawn (using small circles) outside the main plot.

4.3.5   Multiple Graphs

It is often informative to display multiple graphs at the same time in a table format, often referred to as a matrix. This gives an overview of the data from multiple angles. In Figure 4.14, a series of variables have been plotted profiling the frequency distribution for variables in the data set.

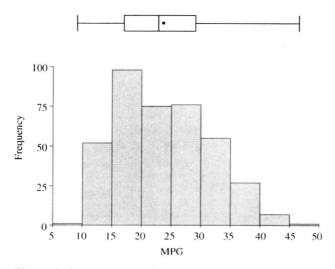

Figure 4.13.   Box plot and histogram side-by-side

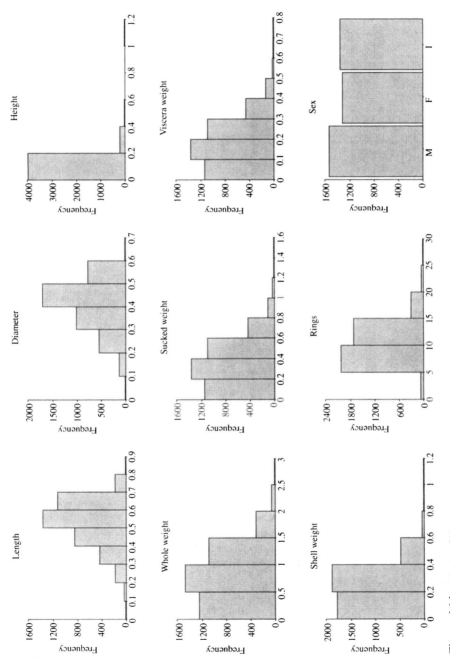

Figure 4.14. Matrix of histograms

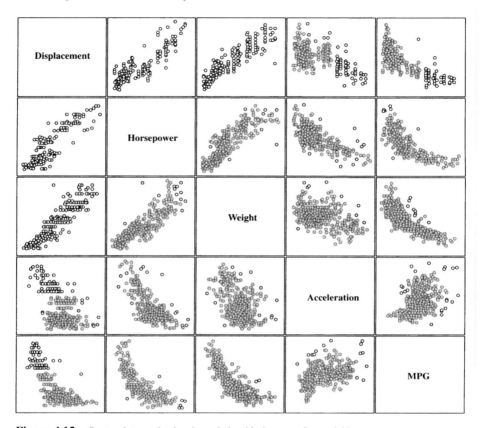

Figure 4.15.   Scatterplot matrix showing relationship between five variables

In Figure 4.15, a series of variables are plotted: **Displacement, Horsepower, Weight, Acceleration,** and **MPG**. This scatterplot matrix shows a series of scatterplots for all pairs of the five variables displayed. The first row shows the relationships between **Displacement** and the four other variables: **Horsepower, Weight, Acceleration,** and **MPG**. The **Displacement** variable is plotted on the *y*-axis for these four graphs. The second row shows the relationship between **Horsepower** and the four other variables. Similarly, the first column shows the relationship between **Displacement** and the four other variables, with **Displacement** plotted on the *x*-axis. Scatterplot matrices are useful to understand key relationships when a data set has many variables.

In Figure 4.16, a set of observations concerning cars have been broken down by year, from 1970 to 1982. Each box plot summarizes the frequency distribution for the variable **MPG** (miles per gallon), for each year. The graph shows how the distribution of car fuel efficiency (**MPG**) has changed over the years.

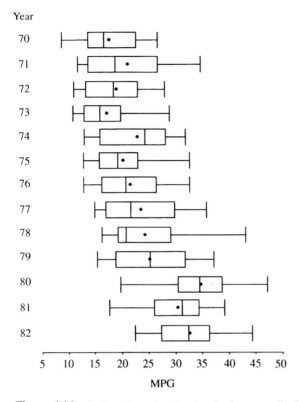

Figure 4.16. Series of box plots showing the frequency distributions over time

Highlighting common subsets of the data can further identify trends in the data set and is illustrated using the automobile example. In Figure 4.17, the shaded area of the graphs are observations where the number of cylinders is 8 (as shown in the top left graph). The other graphs highlight where cars with 8 cylinders can be found on the other frequency distributions. For example, these cars are associated with poor fuel efficiency as shown in the graph in the bottom right (**MPG**). In Figure 4.18, 4-cylinder vehicles are highlighted and it can be seen that the fuel efficiency is generally higher.

4.4 SUMMARY

Table 4.7 summarizes the use of tables and graphs described in this chapter and their use in exploratory data analysis and data mining.

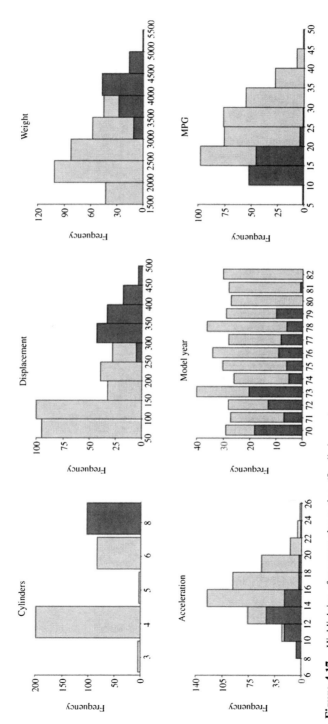

Figure 4.17. Highlighting of common observations (8-cylinder cars)

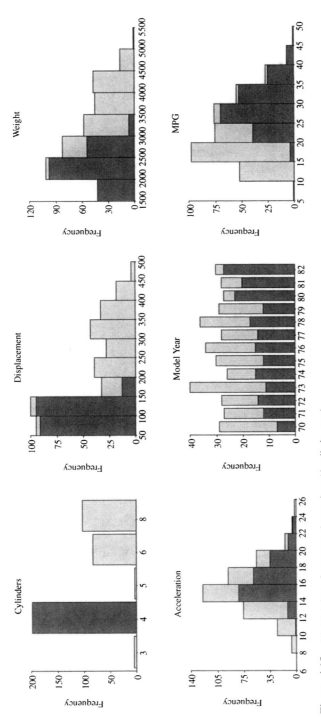

Figure 4.18. Highlighting of common observations (4-cylinder cars)

Table 4.7.   Table summarizing different tables and graphs and their use in analyzing data

	Summary	Data	Uses
Tables	**Raw data table**	Any variables	Showing details of the data
	Contingency table	Two categorical variables	Understanding relationships between categorical variables
	Summary table	Single variable to group observations, other variables to be summarized	Summarizing groups of data
Graphs	**Frequency polygram**	Single variable, any type	Viewing trends, ranges, frequency distribution, and outliers
	Histogram	Single variable, any type	Viewing trends, ranges, frequency distribution, and outliers
	Scatterplot	Two ratio or interval variables	Viewing relationships between continuous variables and outliers
	Box plot	Single ratio, or interval variable	Viewing ranges, frequency distributions, and outliers
	Multiple graphs	Data dependent on individual graph	Viewing multi-dimensional relationships, multi-dimensional summaries, and comparisons

4.5 EXERCISES

Table 4.8 shows a series of retail transactions monitored by the main office of a computer store.

1. Generate a contingency table summarizing the variables **Store** and **Product** category.

2. Generate the following summary tables:

   a. Grouping by **Customer** and the showing a count of the number of observations and the sum of **Sale price ($)** for each row.

   b. Grouping by **Store** and showing a count of the number of observations and the mean **Sale price ($)** for each row.

   c. Grouping by **Product** category and showing a count of the number of observations and the sum of the **Profit ($)** for each row.

Table 4.8.   Retail transaction data set

Customer	Store	Product category	Product description	Sale price ($)	Profit ($)
B.March	New York, NY	Laptop	DR2984	950	190
B.March	New York, NY	Printer	FW288	350	105
B.March	New York, NY	Scanner	BW9338	400	100
J.Bain	New York, NY	Scanner	BW9443	500	125
T.Goss	Washington, DC	Printer	FW199	200	60
T.Goss	Washington, DC	Scanner	BW39339	550	140
L.Nye	New York, NY	Desktop	LR21	600	60
L.Nye	New York, NY	Printer	FW299	300	90
S.Cann	Washington, DC	Desktop	LR21	600	60
E.Sims	Washington, DC	Laptop	DR2983	700	140
P.Judd	New York, NY	Desktop	LR22	700	70
P.Judd	New York, NY	Scanner	FJ3999	200	50
G.Hinton	Washington, DC	Laptop	DR2983	700	140
G.Hinton	Washington, DC	Desktop	LR21	600	60
G.Hinton	Washington, DC	Printer	FW288	350	105
G.Hinton	Washington, DC	Scanner	BW9443	500	125
H.Fu	New York, NY	Desktop	ZX88	450	45
H.Taylor	New York, NY	Scanner	BW9338	400	100

3. Create a histogram of **Sales Price ($)** using the following intervals: 0 to less than 250, 250 to less than 500, 500 to less than 750, 750 to less than 1000.

4. Create a scatterplot showing **Sales price ($)** against **Profit ($)**.

4.6 FURTHER READING

For further reading on communicating information, see Tufte (1990), Tufte (1997), and Tufte (2001). The books also outline good and bad practices in the design of graphs.

The graphs outlined here are essential for exploratory data analysis. There are many alternative charts in addition to the ones described in this chapter. The following web sites describe numerous ways of displaying information graphically: http://www.itl.nist.gov/div898/handbook/eda/eda.htm, http://www.statcan.ca/english/edu/power/toc/contents.htm.

Chapter 5

Statistics

5.1 OVERVIEW

The ability to generate summaries and make general statements about the data, and relationships within the data, is at the heart of exploratory data analysis and data mining methods. In almost every situation we will be making general statements about entire populations, yet we will be using a subset or sample of observations. The distinction between a *population* and a *sample* is important:

- **Population:** A precise definition of all possible outcomes, measurements or values for which inferences will be made about.
- **Sample:** A portion of the population that is representative of the entire population.

Parameters are numbers that characterize a population, whereas *statistics* are numbers that summarize the data collected from a sample of the population. For example, a market researcher asks a portion or a sample of consumers of a particular product, about their preferences, and uses this information to make general statements about all consumers. The entire population, which is of interest, must be defined (i.e. all consumers of the product). Care must be taken in selecting the sample since it must be an unbiased, random sample from the entire population. Using this carefully selected sample, it is possible to make confident statements about the population in any exploratory data analysis or data mining project.

The use of statistical methods can play an important role including:

- **Summarizing the data:** Statistics, not only provide us with methods for summarizing sample data sets, they also allow us to make confident statements about entire populations.
- **Characterizing the data:** Prior to building a predictive model or looking for hidden trends in the data, it is important to characterize the variables and the relationships between them and statistics gives us many tools to accomplish this.

Making Sense of Data: A Practical Guide to Exploratory Data Analysis and Data Mining,
By Glenn J. Myatt
Copyright © 2007 John Wiley & Sons, Inc.

- **Making statements about "hidden" facts:** Once a group of observations, within the data has been defined as interesting through the use of data mining techniques, statistics give us the ability to make confident statements about these groups.

The following chapter describes a number of statistical approaches for making confident decisions. The chapter describes a series of *descriptive statistics* that summarize various attributes of a variable such as the average value or the range of values. *Inferential statistics* cover ways of making confident statements about populations using sample data. Finally, the use of *comparative statistics* allows us to understand relationships between variables.

5.2   DESCRIPTIVE STATISTICS

5.2.1   Overview

Descriptive statistics describe variables in a number of ways. The histogram in Figure 5.1 for the variable **Length** displays the frequency distribution. It can be seen that most of the values are centered around 0.55, with a highest value around 0.85, and a lowest value around 0.05. Most of the values are between 0.3 and 0.7 and the distribution is approximately normal; however, it is slightly skewed.

Descriptive statistics allow us to quantify precisely these descriptions of the data. They calculate different metrics for defining the center of the variable (*central tendency*), they define metrics to understand the range of values (*variation*), and they quantify the shape of the distribution.

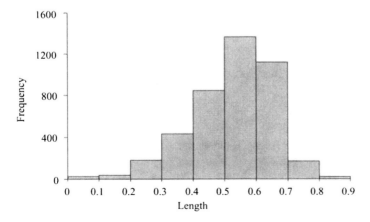

Figure 5.1.   Histogram of variable Length

5.2.2 Central Tendency

Mode

The mode is the most commonly reported value for a particular variable. It is illustrated using the following variable whose values are:

$$3, 4, 5, 6, 7, 7, 7, 8, 8, 9$$

The mode would be the value 7 since there are three occurrences of 7 (more than any other value). It is a useful indication of the central tendency of a variable, since the most frequently occurring value is often towards the center of the variable range.

When there is more than one value with the same (and highest) number of occurrences, either all values are reported or a mid-point is selected. For example, for the following values, both 7 and 8 are reported three times:

$$3, 4, 5, 6, 7, 7, 7, 8, 8, 8, 9$$

The mode may be reported as {7, 8} or 7.5.

Mode provides the only measure of central tendency for variables measured on a nominal scale. The mode can also be calculated for variables measured on the ordinal, interval, and ratio scales.

Median

The median is the middle value of a variable once it has been sorted from low to high. For variables with an even number of values, the mean of the two values closest to the middle is selected (sum the two values and divide by 2).

The following set of values will be used to illustrate:

$$3, 4, 7, 2, 3, 7, 4, 2, 4, 7, 4$$

Before identifying the median, the values must be sorted:

$$2, 2, 3, 3, 4, 4, 4, 4, 7, 7, 7$$

There are 11 values and therefore the sixth value (five values above and five values below) is selected, which is 4:

$$2, 2, 3, 3, 4, \mathit{4}, 4, 4, 7, 7, 7$$

The median can be calculated for variables measured on the ordinal, interval, and ratio scales. It is often the best indication of central tendency for variables measured on the ordinal scale. It is also a good indication of the central value for a variable measured on the interval or ratio scales since, unlike the average, it will not be distorted by any extreme values.

Mean

The mean (also referred to as average) is the most commonly used indication of central tendency for variables measured on the interval or ratio scales. It is defined as the sum of all the values divided by the number of values. For example, for the following set of values:

$$3, 4, 5, 7, 7, 8, 9, 9, 9$$

The sum of all nine values is $(3 + 4 + 5 + 7 + 7 + 8 + 9 + 9 + 9)$ or 61. The sum divided by the number of values is $61 \div 9$ or 6.78.

For a variable representing a sample population (such as x) the mean is commonly referred to as $\bar{x}$. The formula for calculating a mean, where n is the number of observations and x_i is the individual values, is usually written:

$$\bar{x} = \frac{\sum_{i=1}^{n} x_i}{n}$$

Computing the mode, median and mean for a single variable measured on the interval or ratio scale is useful. It is possible to gain an understanding of the shape of the distribution using these values since, if both the mean and median are approximatly the same, the distribution should be fairly symmetrical.

Throughout the book $\bar{x}$ will be used to describe the mean of a sample and μ will be used to describe the population mean.

5.2.3 Variation

Range

The range is a simple measure of the variation for a particular variable. It is calculated as the difference between the highest and lowest values. The following variable will be used to illustrate:

$$2, 3, 4, 6, 7, 7, 8, 9$$

The range is 7 calculated from the highest value (9) minus the lowest value (2).

Range can be used with variables measured on an ordinal, interval or ratio scale.

Quartiles

Quartiles divide a variable into four even segments based on the number of observations. The first quartile (Q1) is at the 25% mark, the second quartile (Q2) is at the 50% mark, and the third quartile (Q3) is at the 75% mark. The calculations for Q1 and Q3 are similar to the calculation of the median. Q2 is the same as the median value. For example, using the following set of values:

$$3, 4, 7, 2, 3, 7, 4, 2, 4, 7, 4$$

The values are sorted:

$$2, 2, 3, 3, 4, 4, 4, 4, 7, 7, 7$$

Next, the median or Q2 is located in the center:

$$2, 2, 3, 3, 4, \boldsymbol{4}, 4, 4, 7, 7, 7$$

If we now look for the center of the first half (shown underlined) or Q1:

$$\underline{2, 2, \boldsymbol{3}, 3, 4,} 4, 4, 4, 7, 7, 7$$

Q1 is recorded as 3. If we now look for the center of the second half (shown underlined) or Q3:

$$2, 2, 3, 3, 4, \boldsymbol{4}, \underline{4, 4, \boldsymbol{7}, 7, 7}$$

Q3 is 7.

Where the boundaries of the quartiles do not fall on a specific value, then the quartile value is calculated based on the two numbers adjacent to the boundary.

The *interquartile range* is defined as the range from Q1 to Q3. In this example it would be $7 - 3$ or 4.

Variance

The variance describes the spread of the data. It is a measure of the deviation of a variable from the mean. For variables that do not represent the entire population, the sample variance formula is:

$$s^2 = \frac{\sum_{i=1}^{n} (x_i - \bar{x})^2}{n - 1}$$

The sample variance is referred to as s^2. The actual value (x_i) minus the mean value ($\bar{x}$) is squared and summed for all values of a variable. This value is divided by the number of observations minus 1 ($n - 1$).

The following example illustrates the calculation of a variance for a particular variable:

$$3, 4, 4, 5, 5, 5, 6, 6, 6, 7, 7, 8, 9$$

Where the mean is:

$$\bar{x} = \frac{3 + 4 + 4 + 5 + 5 + 5 + 6 + 6 + 6 + 7 + 7 + 8 + 9}{13}$$

$$\bar{x} = 5.8$$

Table 5.1 is used to calculate the sum, using the mean value of 5.8. To calculate s^2, we substitute the values from Table 5.1 into the variance formula:

Table 5.1.   Variance intermediate steps

x	$\bar{x}$	$(x_i - \bar{x})$	$(x_i - \bar{x})^2$
3	5.8	−2.8	7.84
4	5.8	−1.8	3.24
4	5.8	−1.8	3.24
5	5.8	−0.8	0.64
5	5.8	−0.8	0.64
5	5.8	−0.8	0.64
6	5.8	0.2	0.04
6	5.8	0.2	0.04
6	5.8	0.2	0.04
7	5.8	1.2	1.44
7	5.8	1.2	1.44
8	5.8	2.2	4.84
9	5.8	3.2	10.24
			Sum = 176.88

$$s^2 = \frac{\sum_{i=1}^{n} (x_i - \bar{x})^2}{n - 1}$$

$$s^2 = \frac{176.88}{13 - 1}$$

$$s^2 = 14.74$$

The variance reflects the average squared deviation. It can be calculated from variables measured on the interval or ratio scale.

The population variance is defined as σ^2 and is calculated using the formula:

$$\sigma^2 = \frac{\sum_{i=1}^{n} (x_i - \bar{x})^2}{n}$$

Standard Deviation

The standard deviation (also described as *root mean square*) is the square root of the variance. For a sample population, the formula is:

$$s = \sqrt{\frac{\sum_{i=1}^{n} (x_i - \bar{x})^2}{n - 1}}$$

Where s is the sample standard deviation, x_i is the actual data value, $\bar{x}$ is the mean for the variable and n is the number of observations. For a calculated variance, for example 14.74, the standard deviation is calculated as $\sqrt{14.74}$ or 3.84.

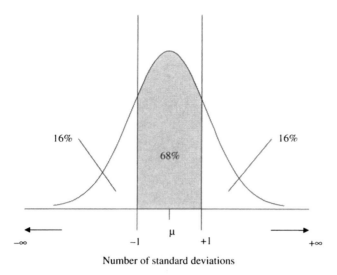

Figure 5.2.   Area under normal distribution from −1 to +1 standard deviations from the mean

The standard deviation is the most widely used expression of the deviation in the range of a variable. The higher the value, the more widely distributed the variable data values are around the mean. Assuming the frequency distribution is approximately normal, about 68% of all observations will fall within one standard deviation of the mean (34% less than and 34% greater than). For example, a variable has a mean value of 45 with a standard deviation value of 6. Approximately 68% of the observations should be in the range 39 to 51 (45 + /− one standard deviation). Figure 5.2 shows that for a normally distributed variable, about 68% of observations fall between −1 and +1 standard deviation. Approximately 95% of all observations fall within two standard deviations of the mean, as shown in Figure 5.3.

Standard deviations can be calculated for variables measured on the interval or ratio scales.

The standard deviation of an entire population will be referred to as σ, which is the square root of the population variance (σ^2).

z-score

A *z-score* represents how far from the mean a particular value is, based on the number of standard deviations. If a *z-score* is calculated for a particular variable, then the *z-score* mean will be zero and each value will reflect the number of standard deviations above or below the mean. Approximately 68% of all observation would be assigned a number between −1 and +1 and approximately 95% of all observations would be assigned a *z-score* between −2 and +2. The following equation is used to calculate a *z-score*:

$$z = \frac{x_i - \bar{x}}{s}$$

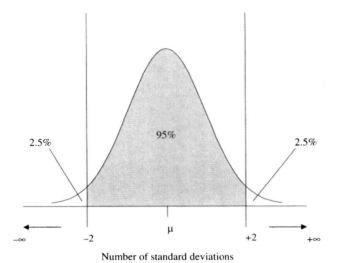

Number of standard deviations

Figure 5.3. Area under the normal distribution from -2 to $+2$ standard deviations from the mean

where x_i is the data value, $\bar{x}$ is the sample mean and s is the standard deviation of the sample. For example, a variable **Age** has values that range from 22 to 97, with a mean of 63.55 and a standard deviation of 13.95. Table 5.2 illustrates a few example calculations for the z-score.

5.2.4 Shape

Skewness

There are methods for quantifying the lack of symmetry or skewness in the distribution of a variable. The formula to calculate skewness, for a variable x, with individual values x_i, with n data points, and a standard deviation of s is:

$$\text{skewness} = \frac{\sum_{i=1}^{n}(x_i - \bar{x})^3}{(n-1)s^3}$$

Table 5.2. Examples of z-score calculation

Age (x_i)	$\bar{x}$	$x_i - \bar{x}$	$z = \frac{x_i - \bar{x}}{s}$
35	63.55	-28.55	-2.05
57	63.55	-6.55	-0.47
63	63.55	-0.55	-0.04
69	63.55	5.45	0.39
81	63.55	17.45	1.25

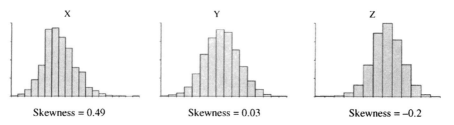

X	Y	Z
Skewness = 0.49	Skewness = 0.03	Skewness = –0.2

Figure 5.4.   Examples illustrating skewness

s^3 is the standard deviation cubed or $s \times s \times s$. A skewness value of zero indicates that the distribution is symmetrical. If the right tail is longer than the left tail then the value is positive and if the left tail is longer than the right tail then the skewness score is negative. Figure 5.4 shows example skewness values for three variables.

Kurtosis

In addition to the symmetry of the distribution, the type of peak that the distribution has, is important to consider. This measurement is defined as kurtosis. The following formula can be used for calculating kurtosis for a variable x, with x_i representing the individual values, with n data points and a standard deviation of s:

$$\text{kurtosis} = \frac{\sum_{i=1}^{n}(x_i - \bar{x})^4}{(n-1)s^4}$$

Variables with a pronounced peak toward the mean have a high kurtosis score and variables with a flat peak have a low kurtosis score. Figure 5.5 illustrates kurtosis scores for two variables.

5.2.5   Example

Figure 5.6 presents a series of descriptive statistics for a variable **Age**. In this example, there are four values for **Age** that occur the most (mode): 69, 76, 64 and 63.

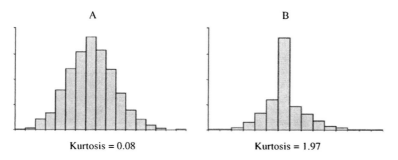

A	B
Kurtosis = 0.08	Kurtosis = 1.97

Figure 5.5.   Examples illustrating kurtosis

Number of observations: 759

Central tendency
Mode: 69, 76, 64, 63
Median: 64
Mean: 63.55

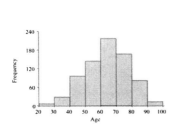

Variation
Minimum: 22
Maximum: 97
Range: 75
Quartiles: 54 (Q1), 64 (Q2), 75 (Q3)
Variance: 194.65
Standard deviation: 13.95

Shape
Skewness: -0.22
Kurtosis: -0.47

Figure 5.6. Descriptive statistics for variable **Age**

The median age is 64 with the mean slightly lower at 63.55. The minimum value is 22 and the maximum value is 97. Half of all observations fall within the range 54–75. The variance is 194.65 and the standard deviation is calculated at 13.95. The distribution is slightly skewed with a longer tail to the left, indicated by the skewness score of -0.22 and the peak is fairly flat indicated by the kurtosis score of -0.47.

5.3 INFERENTIAL STATISTICS

5.3.1 Overview

In almost all situations, we are making statements about populations using data collected from samples. For example, a factory producing packets of sweets believes that there are more than 200 sweets in each packet. To determine a reasonably accurate assessment, it is not necessary to examine every packet produced. Instead an unbiased random sample from this total population could be used.

If this process of selecting a random sample was repeated a number of times, the means from each sample would be different. Different samples will contain different observations and so it is not surprising that the results will change. This is referred to as *sampling error*. If we were to generate many random samples, we might expect that most of the samples would have an average close to the actual mean. We might also expect that there would be a few samples with averages further away from the mean. In fact, the distribution of the mean values follows a normal distribution for sample sizes greater than 30. We will refer to this distribution as the *sampling distribution*, as shown in Figure 5.7.

The sampling distribution is normally distributed because of the *central limit theorem*, which is discussed in the further readings section of the chapter. In fact, the

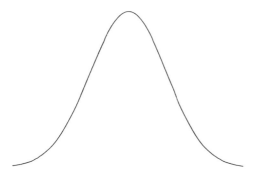

Sampling distribution of $\bar{x}$

Figure 5.7. Sampling distribution for mean values of x

variation of this sampling distribution is dependent on the variation of the variable from which we are now measuring sample means.

We might also expect that increasing the number in each sample would result in more of the sample means being closer to the actual mean. As the sample size increases, the distribution of the means will in fact become narrower, as illustrated in Figure 5.8.

The relationship between the variation of the original variable and the number of observations in the sample to the sampling distribution is summarized in the following formula:

$$\sigma_{\bar{x}} = \frac{\sigma}{\sqrt{n}}$$

As the number of samples increases, the sampling distribution becomes more narrow

$\longrightarrow$

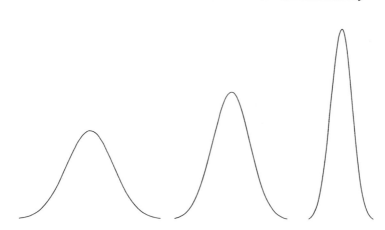

Figure 5.8. Illustration showing that when the number of samples increases, the sampling distribution becomes more narrow

The standard deviation for the distribution of the sample means ($\sigma_{\bar{x}}$) is based on the standard deviation of the population (σ) and the number of observations in the sample (n). As the number of sample observations increases, the standard deviation of the sample means decreases. The standard deviation of the sample means is also called the *standard error of the mean*. Since we rarely have the population standard deviation (σ), the sample standard deviation (s) can be used as an estimate.

We can use this sampling distribution to assess the chance or *probability* that we will see a particular range of average values, which is central to inferential statistics. For example, a sweet manufacturer wishes to make the claim that the average sweets per packet is greater than 200. The manufacturer collects a sample of 500 packets and counts the number of sweets in each of these packets. The average number of sweets per pack is calculated to be 201 with a standard deviation (s) of 12.

We now need to assess the probability that this value is greater than 200 or whether the difference is simply attributable to the sampling error. We can use the sampling distribution to make this assessment. The area under this curve can be used for assessing probabilities. A probability of 1 indicates a certain event and a probability of 0 indicates an event will never happen. Values between these two extremes reflect the relative likelihood of an event happening. The total area under the normal distribution curve is equal to 1. The area between specific *z-score* ranges represents the probability that a value would lie within this range. Therefore, we need to understand where on the normal distribution curve the recorded value lies (see Figure 5.9).

First we calculate the standard error using the sample standard deviation of 12 and the sample size of 500:

$$\sigma_{\bar{x}} = \frac{12}{\sqrt{500}} = 0.54$$

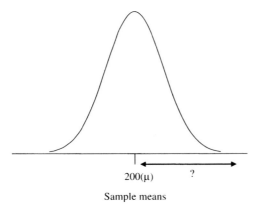

Sample means

Figure 5.9. Determining where the recorded value lies on the sampling distribution

To understand how many standard deviations the value 201 is away from the mean, we must convert the value into a *z-score*:

$$z = \frac{\bar{x} - \mu}{\sigma_{\bar{x}}}$$

where $\bar{x}$ is the mean recorded (201), μ is the population mean in the statement (200) and $\sigma_{\bar{x}}$ is the standard error (0.54). Substituting these values into the formula:

$$z = \frac{201 - 200}{0.54} = 1.85$$

We can now plot the recorded value (converted to a *z-score*) on to the sampling distribution to understand where on this curve the value lies (Figure 5.10). The *z-score* of 1.85 indicates that the 201 value recorded from the sample is higher than the 200 claimed value. The area under the curve to the right of 1.85 can be used to assess the claim. A formal procedure for making these claims will be introduced in this section.

If data is recorded for a categorical variable, instead of examining the average value, we can calculate the proportion of observations with a specific value. For example, a factory producing clothes wishes to understand the number of garments it produces with defects. They use a representative sample of the entire population and record which garments did and did not have defects. To get an overall assessment of the number of defects, a proportion (p) is calculated taking the number of defects and dividing it by the number of observations in the sample. If it is determined that there were 20 defects and the sample size was 200, then the proportion of defects will be $20 \div 200$ or 0.1 (i.e 10%).

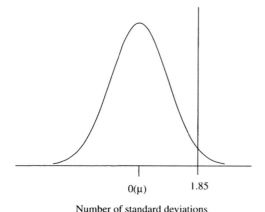

0(μ)   1.85

Number of standard deviations

Figure 5.10. Sampling distribution with the observed value plotted using the *z-score*

In most situations, if we took many different samples and determined the proportions for these samples, the distribution of these proportions would again follow a normal distribution. This normal distribution has a standard deviation (σ_p) which is calculated using the formula:

$$\sigma_p = \sqrt{\frac{p(1-p)}{n}}$$

In this equation n is the sample size and p is the proportion calculated (substituted for the population proportion since it is not usually available). The standard deviation of the proportions is also referred to as the *standard error of proportion*. The sampling distribution of the proportions can be used to estimate a probability that a specific range of proportions would be seen.

In the following sections, we will make use of these standard error calculations and present a number of methods for making statements about data with confidence. The following methods will be discussed:

- **Confidence intervals:** A confidence interval allows us to make statements concerning the likely range that a population parameter (such as the mean) lies within. For example, we may describe the average value falling between 201 and 203 sweets per packet to reflect our level of confidence in the estimate.

- **Hypothesis tests:** A hypothesis test determines whether the data collected supports a specific claim. A hypothesis test can refer to a single group, for example, a hypothesis test may be used to evaluate the claim that the number of sweets per packet is greater than 200. In this example, we are only looking at a single population of packets of sweets. A hypothesis claim can also refer to two groups, for example, to understand if there is a difference in the number of sweets per packet produced by two different machines.

- **Chi-square:** The chi-square test is a statistical test procedure to understand whether a relationship exists between pairs of categorical variables. For example, whether there is a difference in the number of defective garments between three similar factories.

- **One-way analysis of variance:** This test determines whether a relationship exists between three or more group means. For example, if there were more than two machines generating packets of sweets, it would test whether there is a difference between them.

Table 5.3 summarizes the tests discussed in this section.

5.3.2 Confidence Intervals

Overview

A single statistic could be used as an estimate for a population (commonly referred to as a *point estimate*). A single value, however, would not reflect any amount of

Table 5.3.   Summary of inferential statistical tests

	Continuous	Categorical	Number of groups	Number of variables
Confidence intervals	Yes	Yes	1	1
Hypothesis test	Yes	Yes	1 or 2	1
Chi-square	No	Yes	2+	2
One-way analysis of variance	Yes	No	3+	1

confidence in the value. For example, in making an assessment of the average number of sweets per packet we may, based on the number of samples recorded, have a reasonable confidence that this number is between 198 and 202. This range of values is referred to as the *confidence interval*. If a smaller number of samples were collected, we may need to increase the range so that we have confidence that the value lies between, for example, 190 and 210.

The confidence interval is not only dependent on the number of samples collected but is also dependent on the required degree of confidence in the range. If we wish to make a more confident statement, we would have to make the range larger. This required degree of confidence is based on the *confidence level* at which the estimate is to be calculated. The following sections will describe the methods for calculating confidence intervals for continuous and categorical data based on the confidence levels.

Confidence Ranges for Continuous Variables

For continuous variables, the mean is the most common population estimate. For example, using the sweet packet example, the mean would be the sum of all counts divided by the number of packets in the sample. To calculate the confidence interval, we must calculate the mean first. The confidence interval (the range above and below the mean) is dependent on (1) the standard error of the mean and (2) the confidence with which we wish to state the range. The formula for calculating a confidence interval for a large sample (greater than or equal to 30 observations) takes these two factors into consideration:

$$\bar{x} \pm z_C \frac{s}{\sqrt{n}}$$

where $\bar{x}$ is the mean for the sample and $\frac{s}{\sqrt{n}}$ is the standard error of the mean (where s is the standard deviation of the sample and n is the number of observations). The critical *z-score* (z_C) is the number of standard deviations for a given confidence level. To obtain this value, a confidence level needs to be defined. Commonly used, confidence levels include 90%, 95%, and 99%. The critical *z-score* value is calculated by looking at the area under the normal distribution curve at the specified confidence level. As an example, we will use a 95% confidence level, as shown in

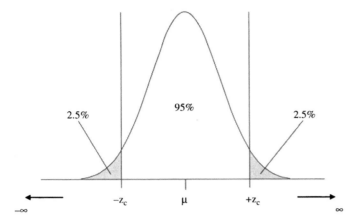

Figure 5.11. Display of the critical *z-score* at a 95% confidence level

Figure 5.11. We need to find the critical *z-score* value where the area in the two shaded extremes totals 5% (2.5% at each shaded region). To look up this *z-score*, we will use the normal distribution table from Appendix A.1. A sample of this table is shown in Figure 5.12. Looking up an area of 2.5% or 0.0250, we see that the corresponding *z-score* is 1.96.

This *z-score* will be used to calculate a confidence interval for a set of 54 observations with a mean value of 33.25 and a standard deviation of 12.26:

$$\bar{x} \pm z_C \frac{s}{\sqrt{n}}$$

$$33.25 \pm 1.96 \frac{12.26}{\sqrt{54}}$$

$$33.25 \pm 3.27$$

Hence, at a 95% confidence level, the confidence interval is from 29.98 to 36.52.

The normal distribution can be used for large sample sizes where the number of observations is greater than or equal to 30. However, for a sample size of less than 30, an alternative distribution is needed: *Student's t-distribution*. This is because the number of observations falls below 30 where we can no longer rely on the normal distribution and instead we must rely on a distribution that has fatter tails. This distribution will result in larger confidence intervals for smaller sample sizes. For sample sizes greater than 30, the t-distribution is similar to the normal distribution and is often used in all situations where the population standard deviation is unknown. The formula for calculating the confidence interval is:

$$\bar{x} \pm t_C \frac{s}{\sqrt{n}}$$

where $\bar{x}$ is the mean of the sample, t_C is the critical *t-value*, s is the standard deviation of the sample, and n is the number of sample observations. This formula

z	.00	.01	.02	.03	.04	.05	.06	.07	.08	.09
0.0	0.5000	0.4960	0.4920	0.4880	0.4841	0.4801	0.4761	0.4721	0.4641	0.4641
0.1	0.4602	0.4562	0.4522	0.4483	0.4443	0.4404	0.4364	0.4325	0.4286	0.4247
0.2	0.4207	0.4168	0.4129	0.4091	0.4052	0.4013	0.3974	0.3936	0.3897	0.3859
0.3	0.3821	0.3783	0.3745	0.3707	0.3669	0.3632	0.3594	0.3557	0.3520	0.3483
0.4	0.3446	0.3409	0.3372	0.3336	0.3300	0.3264	0.3228	0.3192	0.3156	0.3121
0.5	0.3085	0.3050	0.3015	0.2981	0.2946	0.2912	0.2877	0.2843	0.2810	0.2776
0.6	0.2743	0.2709	0.2676	0.2644	0.2611	0.2579	0.2546	0.2514	0.2483	0.2451
0.7	0.2420	0.2389	0.2358	0.2327	0.2297	0.2266	0.2236	0.2207	0.2177	0.2148
0.8	0.2119	0.2090	0.2061	0.2033	0.2005	0.1977	0.1949	0.1922	0.1894	0.1867
0.9	0.1841	0.1814	0.1788	0.1762	0.1736	0.1711	0.1685	0.1660	0.1635	0.1611
1.0	0.1587	0.1363	0.1539	0.1515	0.1492	0.1469	0.1446	0.1423	0.1401	0.1379
1.1	0.1357	0.1335	0.1314	0.1292	0.1271	0.1251	0.1230	0.1210	0.1190	0.1170
1.2	0.1151	0.1131	0.1112	0.1094	0.1075	0.1057	0.1038	0.1020	0.1003	0.0985
1.3	0.0968	0.0951	0.0934	0.0918	0.0901	0.0885	0.0869	0.0853	0.0838	0.0823
1.4	0.0808	0.0793	0.0778	0.0764	0.0749	0.0735	0.0721	0.0708	0.0694	0.0681
1.5	0.0668	0.0655	0.0643	0.0630	0.0618	0.0606	0.0594	0.0582	0.0571	0.0559
1.6	0.0548	0.0537	0.0526	0.0516	0.0505	0.0495	0.0485	0.0475	0.0465	0.0455
1.7	0.0446	0.0436	0.0427	0.0418	0.0409	0.0401	0.0392	0.0384	0.0375	0.0367
1.8	0.0359	0.0351	0.0344	0.0336	0.0329	0.0322	0.0314	0.0307	0.0301	0.0294
1.9	0.0287	0.0281	0.0274	0.0268	0.0262	0.0256	0.0250	0.0244	0.0239	0.0233
2.0	0.0228	0.0222	0.0217	0.0212	0.0207	0.0202	0.0197	0.0192	0.0188	0.0183
2.1	0.0179	0.0174	0.0170	0.0166	0.0162	0.0158	0.0154	0.0150	0.0146	0.0143
2.2	0.0139	0.0136	0.0132	0.0129	0.0125	0.0122	0.0119	0.0116	0.0113	0.0110
2.3	0.0107	0.0104	0.0102	0.0099	0.0096	0.0094	0.0091	0.0089	0.0087	0.0084
2.4	0.0082	0.0080	0.0078	0.0075	0.0073	0.0071	0.0069	0.0068	0.0066	0.0064

An area of 0.0250 has a z-score of 1.96

Adapted with rounding from Table II of R. A. Fisher and F. Yates, *Statistical Tables for Biological, Agricultural and Medical Research*, Sixth Edition, Pearson Education Limited. © 1963 R. A. Fisher and F. Yates

Figure 5.12. Determining the critical z-score from the normal distribution table

can only be applied in situations where the target population approximates a normal distribution.

The value of t_C is calculated using the student's t-distribution table from Appendix A.2. To look-up a *t-value* requires the number of degrees of freedom (*df*) to be specified. The number of degrees of freedom equals the number of observations minus 1. For example, if there were 11 observations, then the number of degrees of freedom will be 10. To look up a critical *t-value* at a confidence level of 95%, the area under the curve right of the critical *t-value* will be 2.5% (0.025). Using the number of degrees of freedom and the area under the curve, it can be seen that the critical *t-value* is 2.228, as shown in Figure 5.13.

In the following example, a set of 11 (*n*) observations was recorded and the mean value was calculated at 23.22 ($\bar{x}$), with a standard deviation of 11.98 (*s*). At a 95% confidence level, the value of t_C is 2.228 and hence the confidence interval is:

$$\bar{x} \pm t_C \frac{s}{\sqrt{n}}$$

$$23.22 \pm 2.228 \frac{11.98}{\sqrt{11}}$$

$$23.22 \pm 8.05$$

Upper tail area

df	0.1	0.05	0.025	0.01	0.005
1	3.078	6.314	12.706	31.821	63.657
2	1.886	2.920	4.303	6.965	9.925
3	1.638	2.353	3.182	4.541	5.841
4	1.533	2.132	2.776	3.747	4.604
5	1.476	2.015	2.571	3.365	4.032
6	1.440	1.943	2.447	3.143	3.707
7	1.415	1.895	2.365	2.998	3.499
8	1.397	1.860	2.306	2.896	3.355
9	1.383	1.833	2.262	2.821	3.250
10	1.372	1.812	2.228	2.764	3.169
11	1.363	1.796	2.201	2.718	3.106
12	1.356	1.782	2.179	2.681	3.055
13	1.350	1.771	2.160	2.650	3.012
14	1.345	1.761	2.145	2.624	2.977
15	1.341	1.753	2.131	2.602	2.947

An area of 0.025 with 10 degrees of freedom (*df*) has a *t-value* of 2.228

Adapted from Table III of R. A. Fisher and F. Yates, *Statistical Tables for Biological, Agricultural and Medical Research*, Sixth Edition, Pearson Education Limited, © 1963 R. A. Fisher and F. Yates

Figure 5.13. Determining the critical value of *t* using the t-distribution

Hence at a 95% confidence level, the confidence interval is from 15.17 to 31.27.

Confidence Ranges for Categorical Variables

When handling categorical variables, the proportion with a given outcome is often used to summarize the variable. This equals the outcome's size divided by the sample size. For example, a factory may be interested in the proportion of units produced with errors. To make this assessment, a sample of 300 units are tested for errors, and it is determined that 45 contain a problem. The proportion of units in the sample with errors is 45/300 or 0.15. To make a statement about the population as a whole, it is important to indicate the confidence interval. Again, this is based on (1) the standard error of the proportion and (2) the confidence level with which we wish to state the range. In this example, we will use a confidence level of 95%. Based on this information, the following equation can be used to determine the range:

$$p \pm z_C \sqrt{\frac{p(1-p)}{n}}$$

where p is the proportion with a given outcome, n is the sample size and z_C is the critical z-score. $\sqrt{\frac{p(1-p)}{n}}$ is the standard error of proportion. For this example, p is 45 divided by 300 or 0.15 (15%) and n is 300. The critical z-score is computed based on the confidence level and is determined using the area under a normally distributed curve as described earlier. Given a 95% confidence level, the area under the upper and lower tails marked in gray should be 5% or 0.05 (see Figure 5.11). Hence the area under the lower tail should be 0.025 and the area under the upper tail should be 0.025. The z-score can be calculated from the tables in Appendix A.1. The critical z-score for a 95% confidence level is 1.96. Substituting these values into the equation:

$$p \pm z_C \sqrt{\frac{p(1-p)}{n}} \quad \text{or} \quad 0.15 \pm 1.96 \sqrt{\frac{0.15 \times (1-0.15)}{300}}$$

$$0.15 \pm 0.04$$

It can be inferred that between 11% and 19% of units will have faults with 95% confidence.

5.3.3 Hypothesis Tests

Overview

In this example, a clothing manufacturer wishes to make a claim concerning the number of garments it creates with no defects. It believes that less than 5% contain a defect. To examine every garment produced would be too costly and so they decided to collect 500 garments, selected randomly. Each garment is examined and it is recorded whether the garment has a defect or not. After the data was collected, it was calculated that 4.7% of garments had defects. Since the data was collected from a sample, this number alone is not sufficient to make any claim because of the potential for sampling errors. Knowing that they would not be able to make a claim with 100% confidence, they would be satisfied with a 95% confidence rate, that is, 95 times out of 100 they would be correct. The sampling distribution, described earlier, can be used to understand the minimum number of defective garments to make the claim with a 95% confidence. This point should be plotted on the sampling distribution. If the 4.7% value is now plotted on the sampling distribution (by converting it to a *z-score*), it should now be possible to understand whether it is sufficiently low to make the statement at a 95% confidence. If it is not, then the manufacturer would not be able to make any claims about the number of defective garments. The following describes a formal procedure for making claims or *hypothesis* using data.

A hypothesis is a statement or claim made about an entire population. For example:

- The average time to process a passport is 12 days
- More than eight out of ten dog owners prefer a certain brand (brand X) of dog food.

A hypothesis test determines whether you have enough data to reject the claim (and accept the alternative) or whether you do not have enough data to reject the claim. To define a hypothesis, two statements are made:

- **Null hypothesis** (H_0): This is a claim that a particular population parameter (e.g. mean) equals a specific value. For example, the average time to process a passport equals 12 days or the proportion of dog owners that prefer brand X is 0.8 (or 80%). A hypothesis test will either reject or not reject the null hypothesis using the collected data.
- **Alternative hypothesis** (H_a): This is the conclusion that we would be interested in reaching if the null hypothesis is rejected. Another name that is used to describe the alternative hypothesis is the *research hypothesis* as it is often the conclusion that the researcher is interested in reaching. There are three options: not equal to, greater than or less than. For example, dog owners' preference for brand X dog food is more than 0.8 (or 80%) or the passport processing time is either greater than or less than 12 days so that the alternative hypothesis is defined as not equal to 12 days.

To illustrate the null and alternative hypothesis, we will use the two cases described above:

- **Claim:** The average time to process a passport is 12 days

$$H_0 : \mu = 12$$
$$H_a : \mu \neq 12$$

where μ is the claimed average number of days to process a passport.

- **Claim:** More than eight out of ten dog owners prefer a certain brand (brand X) of dog food

$$H_0 : \pi = 0.8$$
$$H_a : \pi > 0.8$$

where π is the claimed proportion of dog owners preferring brand X dog food. The alternative hypothesis is that the proportion of dog owner who prefer brand X is greater than 0.8 and we would be interested in reaching this conclusion if the null hypothesis was rejected.

Hypothesis Assessment

Before a hypothesis test is performed it is necessary to set a value at which H_0 should be rejected. Since we are dealing with a sample of the population, the hypothesis test may be wrong. We can, however, minimize the chance of an error by specifying a confidence level that reflects the chance of an error. For example, setting a confidence level at 90% means that we would expect 1 in 10 results to be incorrect, whereas setting a confidence level at 99% we would expect 1 in 100 incorrect results. A typical value is 95% confidence; however, values between 90%–99% are often used. This is the point at which H_0 will be rejected. This confidence level is usually described by the term α, which is 100 minus the confidence percentage level, divided by 100. For example, a 95% confidence level has $\alpha = 0.05$ and a 99% confidence level has $\alpha = 0.01$.

Once the null hypothesis and the alternative hypothesis have been described, it is now possible to assess the hypotheses using the data collected. First, the statistic of interest from the sample is calculated. Next, a hypothesis test will look at the difference between the value claimed in the hypothesis statement and the calculated sample statistic. For large sample sets (greater than or equal to 30 observations), identifying where the hypothesis test result is located on the normal distribution curve of the sampling distribution, will determine whether the null hypothesis is rejected.

For example, the following graph (Figure 5.14) shows a two-tail test used in situations where the alternative hypothesis is expressed as not equal. In this example, we use a confidence level where $\alpha = 0.05$. The graph shows the standard normal distribution with the null hypothesis parameter shown in the center of the graph (μ_{H_0}). If the hypothesis test score is within the "do not reject H_0" region, then there is not enough evidence to reject the null hypothesis. If the hypothesis test score is in the "reject H_0" region, then the null hypothesis is rejected. The value of z_c is determined from the normal distribution table in Appendix A.1. Since this is a two-tail test, the sum of the area in the two tails should equal 5%, as shown in Figure 5.14.

If the alternative hypothesis is smaller than, then you reject the null hypothesis only if it falls in the left "reject H_0" region. If the alternative hypothesis is greater than, then you reject the null hypothesis if it falls in the right "reject H_0" region. For

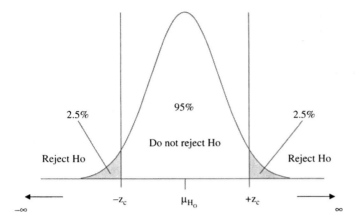

Figure 5.14. Illustration of the two-tail reject/do not reject region when $\alpha = 0.05$

example, if the alternative hypothesis is H_a: $\mu < \mu_{H_0}$ and $\alpha = 0.05$, then we would reject the null hypothesis if the hypothesis test results has a *z-score* to the left of the critical value of z (z_c). The value of z_c is determined from the normal distribution table in Appendix A.1. Since this is a one-tail test (when smaller than or greater than is in the alternative hypothesis) the single area should equal 5%, as shown in Figure 5.15.

Calculating p-Values

A hypothesis test is usually converted into a *p-value*. A *p-value* is the probability of getting the recorded value or a more extreme value. It is a measure of the likelihood

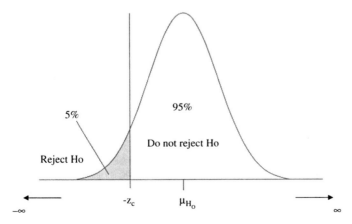

Figure 5.15. Illustration of the one-tail reject/do not reject region when $\alpha = 0.05$

of the result given the null hypothesis is true or the statistical significance of the claim. To calculate a *p-value*, use the score calculated using the hypothesis test (described in the next sections) and look up the score on the standardized normal distribution. For example, a hypothesis score of 2.21 corresponds to a value of 0.0136 (see Figure 5.16). Where the alternative hypothesis is not equal, then this value is doubled.

p-values range from 0 to 1. Where the *p-value* is less than α, the null hypothesis is rejected. When the *p-value* is greater than α, the null hypothesis is not rejected. For example, if α was set to 0.05, then a *p-value* of 0.0136 would mean we would reject the null hypothesis.

Where a sample size is small (less than 30), a student's *t*-distribution should be used instead of the standard normal distribution in calculating a *p-value* (see appendix A.2).

Hypothesis Test: Single Group, Continuous Data

To test the claim that the average time to process a passport is 12 days, the following null and alternative hypothesis were defined:

$$H_0 : \mu = 12$$
$$H_a : \mu \neq 12$$

where μ is the claimed average number of days to process a passport.

To test the hypothesis that the number of days to process a passport is 12 (μ), 45 passport applications were randomly selected and timed ($n = 45$). The average time to process the passport application was 12.1 ($\bar{x}$) and the standard deviation was 0.23 (s) and α was set to 0.05. To calculate the hypothesis test, the following formula will be used:

$$z = \frac{\bar{x} - \mu}{\frac{s}{\sqrt{n}}}$$

This formula uses the difference between the actual mean and the null hypothesis mean, divided by the standard error of the mean. In this example:

$$z = \frac{12.1 - 12.0}{\frac{0.23}{\sqrt{45}}}$$
$$z = 2.9$$

for a value of $\alpha = 0.05$, the critical value of z (z_c) would be 1.96. This is where the area under each extreme would equal 2.5%. Since the *z-score* of 2.9 is greater than 1.96, we reject the null hypothesis and make a statement that the average number of days to process a passport is not 12 days. To calculate a *p-value*, we look-up the calculated hypothesis score of 2.9 in the normal distribution table and this value is 0.0019. Since this hypothesis is two-sided, we double this value to obtain a *p-value* of 0.0038.

z	.00	.01	.02	.03	.04	.05	.06	.07	.08	.09
0.0	0.5000	0.4960	0.4920	0.4880	0.4841	0.4801	0.4761	0.4721	0.4681	0.4641
0.1	0.4602	0.4562	0.4522	0.4483	0.4443	0.4404	0.4364	0.4325	0.4286	0.4247
0.2	0.4207	0.4168	0.4129	0.4091	0.4052	0.4013	0.3974	0.3936	0.3897	0.3859
0.3	0.3821	0.3783	0.3745	0.3707	0.3669	0.3632	0.3594	0.3557	0.3520	0.3483
0.4	0.3446	0.3409	0.3372	0.3336	0.3300	0.3264	0.3228	0.3192	0.3156	0.3121
0.5	0.3085	0.3050	0.3015	0.2981	0.2946	0.2912	0.2877	0.2843	0.2810	0.2776
0.6	0.2743	0.2709	0.2676	0.2644	0.2611	0.2579	0.2546	0.2514	0.2483	0.2451
0.7	0.2420	0.2389	0.2358	0.2327	0.2297	0.2266	0.2236	0.2207	0.2177	0.2148
0.8	0.2119	0.2090	0.2061	0.2033	0.2005	0.1977	0.1949	0.1922	0.1894	0.1867
0.9	0.1841	0.1814	0.1788	0.1762	0.1736	0.1711	0.1685	0.1660	0.1635	0.1611
1.0	0.1587	0.1563	0.1539	0.1515	0.1492	0.1469	0.1446	0.1423	0.1401	0.1379
1.1	0.1357	0.1335	0.1314	0.1292	0.1271	0.1251	0.1230	0.1210	0.1190	0.1170
1.2	0.1151	0.1131	0.1112	0.1094	0.1075	0.1057	0.1038	0.1020	0.1003	0.0985
1.3	0.0968	0.0951	0.0934	0.0918	0.0901	0.0885	0.0869	0.0853	0.0838	0.0823
1.4	0.0808	0.0793	0.0778	0.0764	0.0749	0.0735	0.0721	0.0708	0.0694	0.0681
1.5	0.0668	0.0655	0.0643	0.0630	0.0618	0.0606	0.0594	0.0582	0.0571	0.0559
1.6	0.0548	0.0537	0.0526	0.0516	0.0505	0.0495	0.0485	0.0475	0.0465	0.0455
1.7	0.0446	0.0436	0.0427	0.0418	0.0409	0.0401	0.0392	0.0384	0.0375	0.0367
1.8	0.0359	0.0351	0.0344	0.0336	0.0329	0.0322	0.0314	0.0307	0.0301	0.0294
1.9	0.0287	0.0281	0.0274	0.0268	0.0262	0.0256	0.0250	0.0244	0.0239	0.0233
2.0	0.0228	0.0222	0.0217	0.0212	0.0207	0.0202	0.0197	0.0192	0.0188	0.0183
2.1	0.0179	0.0174	0.0170	0.0166	0.0162	0.0158	0.0154	0.0150	0.0146	0.0143
2.2	0.0139	0.0136	0.0132	0.0129	0.0125	0.0122	0.0119	0.0116	0.0113	0.0110
2.3	0.0107	0.0104	0.0102	0.0099	0.0096	0.0094	0.0091	0.0089	0.0087	0.0084
2.4	0.0082	0.0080	0.0078	0.0075	0.0073	0.0071	0.0069	0.0068	0.0066	0.0064

Looking up z-score of 2.21
corresponds to an area of 0.0136

Adapted with rounding from Table II of R. A. Fisher and F. Yates, *Statistical Tables for Biological, Agricultural and Medical Research*, Sixth Edition, Pearson Education Limited, © 1963 R. A. Fisher and F. Yates

Figure 5.16. Looking up value from normal distribution table

Hypothesis Test: Single Group, Categorical Data

To test the claim that more than eight out of ten dog owners prefer a certain brand (brand X) of dog food, the following null and alternative hypothesis were defined:

$$H_0 : \pi = 0.8$$
$$H_a : \pi > 0.8$$

where π is the claimed proportion of dog owners preferring brand X dog food.

To test this hypothesis, 40 random dog owners ($n = 40$) were questioned and the proportion that responded that they preferred brand X was 33 out of 40 or 0.825 (p). The proportion in the null hypothesis was 0.8 (π_0) and α was set to 0.05. To calculate the hypothesis test (z), the following formula is used:

$$z = \frac{p - \pi_0}{\sqrt{\frac{\pi_0(1 - \pi_0)}{n}}}$$

This is the difference between the value stated in the null hypothesis and the recorded sample divided by the standard error of proportions. In this example,

$$z = \frac{0.825 - 0.8}{\sqrt{\frac{0.8(1 - 0.8)}{40}}}$$

$$z = 0.395$$

The critical z-score when $\alpha = 0.05$ is 1.65, which is greater than the hypothesis test score. Looking up 0.395 on the standardized normal distribution, we get a p-value of 0.3446. Since the p-value is greater than α, we do not reject the null hypothesis. In this case, we cannot make the claim that more than 80% of dog owners prefer brand X.

Hypothesis Test: Two Groups, Continuous Data

In this example, the following claim is to be tested:

Claim: The average fuel efficiency for 4-cylinder vehicles is greater than the average fuel efficiency for 6-cylinder vehicles.

To test this claim the null and alternative hypothesis are defined:

$$H_0 : \mu_1 = \mu_2$$
$$H_a : \mu_1 > \mu_2$$

where μ_1 is the average fuel efficiency for a population of 4-cylinder vehicles and μ_2 is the average fuel efficiency for a population of 6-cylinder vehicles.

Two groups of cars were randomly selected, one group with four cylinders and one group with six cylinders. The fuel efficiency of each car is collected. The first group is a set of 24 4-cylinder cars (n_1) with an average fuel efficiency (in miles per gallon) of 25.85 ($\bar{x}_1$), and a variance of 50.43 (s_1^2). The second group is a collection of 27 6-cylinder cars (n_2) with an average fuel efficiency (in miles per gallon) of 23.15 ($\bar{x}_2$), and a variance of 48.71 (s_2^2). α is set to 0.05 in this example.

The null hypothesis states that there is no difference between the mean of 4-cylinder cars (μ_1) compared to the mean of 6-cylinder cars (μ_2). The alternative hypothesis states that 4-cylinder vehicles have greater fuel efficiency than 6-cylinder vehicles.

Since the group sizes are less than 30, the following formula will be used:

$$t = \frac{(\bar{x}_1 - \bar{x}_2) - (\mu_1 - \mu_2)}{s_P \sqrt{\frac{1}{n_1} + \frac{1}{n_2}}}$$

where s_P is the pooled standard deviation and can be calculated from s_P^2 (the pooled variance):

$$s_P^2 = \frac{(n_1 - 1)s_1^2 + (n_2 - 1)s_2^2}{(n_1 - 1) + (n_2 - 1)}$$

In the further readings section of this chapter, a number of references describe how these formulas were obtained.

In this example:

$$s_P^2 = \frac{(24 - 1)50.43 + (27 - 1)48.71}{(24 - 1) + (27 - 1)}$$

$$s_P^2 = 49.52$$

Since the null hypothesis states that $\mu_1 = \mu_2$, $\mu_1 - \mu_2 = 0$

$$t = \frac{(25.85 - 23.15) - (0)}{\sqrt{49.52}\sqrt{\frac{1}{24} + \frac{1}{27}}}$$

$$t = 1.37$$

In this example, the number of degrees of freedom is 49 ($n_1 + n_2 - 2$), and the critical value for t is approximately 1.68 (from Appendix A.2). Since the *p-value* is just less than 0.1, we do not reject the null hypothesis. The formulas used in this example can be applied when there are less than 30 observations in either group and when the population is normally distributed. In situations where the number of observations in both groups is greater than 30, the following equation can be used:

$$z = \frac{(\bar{x}_1 - \bar{x}_2) - (\mu_1 - \mu_2)}{\sqrt{\frac{s_1^2}{n_1} + \frac{s_2^2}{n_2}}}$$

where s_1^2 is the variance of the first group and s_2^2 is the variance of the second group.

Hypothesis Test: Two Groups, Categorical Data

In the following example, a claim is made concerning the efficacy of a new drug used to treat strokes:

Claim: A new drug reduces the number of strokes.

To test this claim the following null and alternative hypothesis are defined:

$$H_0 : \pi_1 = \pi_2$$
$$H_a : \pi_1 < \pi_2$$

where π_1 is the proportion of the population with strokes taking the new medicine and π_2 is the proportion of the population with strokes taking the placebo.

Two groups of patients were randomly selected and studied. One of the groups takes a placebo (a sugar pill) and the other group takes the new medicine. The number of strokes for each patient group is recorded. In this situation, the hypothesis test is based on the difference in the proportion of strokes between the two populations. There were 10,004 patients in the first group who took the medicine (n_1) and of these 213 had strokes (X_1). There were 10,013 patients in group 2 that did not take the medicine and took a placebo instead (n_2) and in this group 342 patients had a stroke (X_2). The results of the study are shown in Table 5.4.

Overall, the two groups are examined together to understand the total proportion of patients that had strokes:

$$p = \frac{X_1 + X_2}{n_1 + n_2}$$

$$p = \frac{213 + 342}{10004 + 10013}$$

$$p = 0.0277$$

The proportion of the first group (that takes the medicine) that has strokes is:

$$p_1 = \frac{X_1}{n_1}$$

$$p_1 = \frac{213}{10004}$$

$$p_1 = 0.0213$$

Table 5.4.   Contingency table indicating the results of a medical test

	Takes medicine	Takes placebo	Total
Has strokes	213	342	555
No strokes	9,791	9,671	19,462
Totals	10,004	10,013	20,017

The proportion of the second group (that takes the placebo) that has strokes is:

$$p_2 = \frac{X_2}{n_2}$$

$$p_2 = \frac{342}{10013}$$

$$p_2 = 0.0342$$

The null hypothesis states that there is no difference in the proportion of strokes between the group taking the medicine (π_1) compared to the group not taking the medicine (π_2). To calculate a hypothesis test, the following equation will be used:

$$z = \frac{(p_1 - p_2) - (\pi_1 - \pi_2)}{\sqrt{p(1-p)(\frac{1}{n_1} + \frac{1}{n_2})}}$$

For more information on how this formula was obtained, see the further reading section of this chapter.

In this hypothesis test, ($\pi_1 - \pi_2$) is equal to 0 since there should be no difference according to the null hypothesis.

$$z = \frac{(0.0213 - 0.0342) - 0}{\sqrt{0.0278(1 - 0.0278)(\frac{1}{10004} + \frac{1}{10013})}}$$

$$z = \frac{-0.0129}{0.00232}$$

$$z = -5.54$$

To calculate a *p-value* based on this hypothesis test, we look up this score in the normal distribution table (Appendix A.1) and it is virtually 0, hence we reject the null hypothesis and conclude the number of strokes for the group taking the medicine is lower than the group that does not take the medicine.

Paired Test

In this widely quoted example, the following claim is made:

Claim: There is no difference in the wear of shoes made from material X compared to shoes made from material Y.

To test this claim, the null and alternative hypothesis are set up:

$$H_0 : \mu_D = 0$$
$$H_a : \mu_D \neq 0$$

where μ_D is the difference between the wear of shoes made with material X and the wear of shoes made with material Y.

To test the hypothesis, 10 boys wore a shoe made with material X on one foot and a shoe made with material Y on the other and the feet were randomized. The amount of wear for each material was recorded. A 90% confidence level is required.

The average difference is 0.41 ($\bar{D}$), and the number of standard deviations is 0.386 (s_D) for this difference. Since the number of observations is small, we will use the t-distribution to assess the hypothesis. The following formula is used:

$$t = \frac{\bar{D} - \mu_D}{\frac{s_D}{\sqrt{n}}}$$

$$t = \frac{0.41 - 0}{\frac{0.386}{\sqrt{10}}}$$

$$t = 3.36$$

To calculate a *p-value* based on this hypothesis test, we look up this score in the t-distribution table (Appendix A.2), where the number of degrees of freedom is 9 ($n - 1$). It is just less than 0.01, hence we reject the null hypothesis and conclude that there is a difference.

Errors

Since a hypothesis test is based on a sample and samples vary, there exists the possibility for errors. There are two potential errors and these are described as:

- **Type I Error:** In this situation the null hypothesis is rejected when it really should not be. These errors are minimized by setting the value of α low.
- **Type II Error:** In this situation the null hypothesis is not rejected when it should have been. These errors are minimized by increasing the number of observations in the sample.

5.3.4   Chi-Square

The chi-square test is a hypothesis test to use with variables measured on a nominal or ordinal scale. It allows an analysis of whether there is a relationship between two categorical variables. As with other hypothesis tests, it is necessary to state a null and alternative hypothesis. Generally, these hypothesis statements look like:

H_0: There is no relationship

H_a: There is a relationship

Using Table 5.5, we will look at whether a relationship exists between where a consumer lives (represented by a zip code) and the brand of washing powder they buy (brand X, brand Y, and brand Z).

Table 5.5.   Contingency table of observed purchases

		Washing powder brand			
		Brand X	Brand Y	Brand Z	
Zip code	**43221**	5,521	4,597	4,642	**14,760**
	43029	4,522	4,716	5,047	**14,285**
	43212	4,424	5,124	4,784	**14,332**
		14,467	**14,437**	**14,473**	**43,377**

The chi-square test compares the observed frequencies with the expected frequencies. The expected frequencies are calculated using the following formula:

$$E_{r,c} = \frac{r \times c}{n}$$

where $E_{r,c}$ is the expected frequency for a particular cell in the table, r is the row count, c is the column count and n is the total observations in the sample.

For example, to calculate the expected frequency for the cell where the washing powder is brand X and the zip code is 43221 would be:

$$E_{Brand\_X,43221} = \frac{14,760 \times 14,467}{43,377}$$

$$E_{Brand\_X,43221} = 4,923$$

Table 5.6 shows the entire table with the expected frequency count (replacing the observed count).

The chi-square test (χ^2) is computed with the following equation:

$$\chi^2 = \sum_{i=1}^{k} \frac{(O_i - E_i)^2}{E_i}$$

where k is the number of all categories, O_i is the observed cell frequency and E_i is the expected cell frequency. Table 5.7 shows the computed χ^2 for this example.

There is a critical value at which the null hypothesis is rejected (χ_c^2). This value is found using the chi-square table in Appendix A.3. The value is dependent on the

Table 5.6.   Contingency table of expected purchases

		Washing powder brand			
		Brand X	Brand Y	Brand Z	
Zip code	**43221**	4,923	4,913	4,925	**14,760**
	43026	4,764	4,754	4,766	**14,285**
	43212	4,780	4,770	4,782	**14,332**
		14,467	**14,437**	**14,473**	**43,377**

Table 5.7. Calculation of chi-square

k	Category	Observed (O)	Expected (E)	$(O - E)^2/E$
1	r = Brand X, c = 43221	5,521	4,923	72.6
2	r = Brand Y, c = 43221	4,597	4,913	20.3
3	r = Brand Z, c = 43221	4,642	4,925	16.3
4	r = Brand X, c = 43026	4,522	4,764	12.3
5	r = Brand Y, c = 43026	4,716	4,754	0.3
6	r = Brand Z, c = 43026	5,047	4,766	16.6
7	r = Brand X, c = 43212	4,424	4,780	26.5
8	r = Brand Y, c = 43212	5,124	4,770	26.3
9	r = Brand Z, c = 43212	4,784	4,782	0.0008
				Sum = **191.2**

degrees of freedom (df), which is calculated:

$$df = (r - 1) \times (c - 1)$$

For example, the number of degrees of freedom for this example is $(3 - 1) \times (3 - 1)$ which is 4. Looking up the critical value, for $df = 4$ and $\alpha = 0.05$, the critical value is 9.488 as shown in Figure 5.17. Since 9.488 is less than the calculated chi-square value of 191.2, we reject the null hypothesis and state that there is a relationship between zip codes and brands of washing powder. The chi-square test will tell you if a relationship exists; however, it does not tell you what sort of relationship it is.

5.3.5   One-Way Analysis of Variance

Overview

The following section reviews a technique called one-way analysis of variance that compares the means from three or more different groups. The test determines whether there is a difference between the groups. This method can be applied to cases where the groups are independent and random, the distributions are normal, and the populations have similar variances. For example, an on-line computer retail company has call centers in four different locations. These call centers are approximately the same size and handle a certain number of calls each day. An analysis of the different call centers based on the average number of calls processed each day is required. Table 5.8 illustrates the daily calls serviced.

As with other hypothesis tests, it is necessary to state a null and alternative hypothesis. Generally, the hypothesis statement will look like:

H_0: The sample means are equal

H_a: The sample means are not equal

To determine whether there is a difference or not between the means or whether the difference is due to random variation, we must perform a hypothesis test. This test

| | | | | | | | Probability | | | | | | | |
df	0.99	0.98	0.95	0.90	0.80	0.70	0.50	0.30	0.20	0.10	0.05	0.02	0.01	0.001
1	0.0^3157	0.0^3628	0.00393	0.0158	0.0642	0.148	0.455	1.074	1.642	2.706	3.841	5.412	6.635	10.827
2	0.0201	0.0404	0.103	0.211	0.446	0.713	1.386	2.408	3.219	4.605	5.991	7.824	9.210	13.815
3	0.115	0.185	0.352	0.584	1.005	1.424	2.366	3.665	4.642	6.251	7.815	9.837	11.345	16.266
4	0.297	0.429	0.711	1.064	1.649	2.195	3.357	4.878	5.989	7.779	9.488	11.668	13.277	18.467
5	0.554	0.752	1.145	1.610	2.343	3.000	4.351	6.064	7.289	9.236	11.070	13.388	15.086	20.515
6	0.872	1.134	1.635	2.204	3.070	3.828	5.348	7.231	8.558	10.645	12.592	15.033	16.812	22.457
7	1.239	1.564	2.167	2.833	3.822	4.671	6.346	8.383	9.803	12.017	14.067	16.622	18.475	24.322
8	1.646	2.032	2.733	3.490	4.594	5.527	7.344	9.524	11.030	13.362	15.507	18.168	20.090	26.125
9	2.088	2.532	3.325	4.168	5.380	6.393	8.343	10.656	12.242	14.684	16.919	19.679	21.666	27.877
10	2.558	3.059	3.940	4.865	6.179	7.267	9.342	11.781	13.442	15.987	18.307	21.161	23.209	29.588
11	3.053	3.609	4.575	5.578	6.989	8.148	10.341	12.899	14.631	17.275	19.675	22.618	24.725	31.264
12	3.571	4.178	5.226	6.304	7.807	9.034	11.340	14.011	15.812	18.549	21.026	24.054	26.217	32.909
13	4.107	4.765	5.892	7.042	8.634	9.926	12.340	15.119	16.985	19.812	22.362	25.472	27.688	34.528
14	4.660	5.368	6.571	7.790	9.467	10.821	13.339	16.222	18.151	21.064	23.685	26.873	29.141	36.123
15	5.229	5.985	7.261	8.547	10.307	11.721	14.339	17.322	19.311	22.307	24.996	28.259	30.578	37.697

Looking up critical chi-square value, for $df = 4$ and $\alpha = 0.05$

Adapted from Table IV of R. A. Fisher and F. Yates, *Statistical Tables for Biological, Agricultural and Medical Research*, sixth Edition, Pearson Education Limited, © 1963 R. A. Fisher and F. Yates

1963 R. A. Fisher and F. Yates

Figure 5.17. Looking up the critical chi-square value

Table 5.8. Calls processed by different call centers

Call center A	Call center B	Call center C	Call center D
136	124	142	149
145	131	145	157
139	128	139	154
132	130	145	155
141	129	143	151
143	135	141	156
138	132	138	
139		146	

will look at both the variation within the groups and the variation between the groups. The test has the following steps:

1. Calculate group means and standard deviations
2. Determine the within group variation
3. Determine the between group variation
4. Determine the F-statistic, using the within and between group variation
5. Test the significance of the F-statistic

The following sections describe these steps in detail:

Calculate Group Means and Variances

In Table 5.9, for each call center a count along with the mean and variance has been calculated. In addition, the total number of groups is listed ($k = 4$) and the total number of observations ($N = 29$). In addition, an average of the means ($\bar{\bar{x}} = 140.8$) is calculated by taking each mean value for each call center and dividing it by the number of groups:

$$\bar{\bar{x}} = \frac{139.1 + 129.9 + 142.4 + 153.7}{4} = 141.3$$

Determine the Within Group Variation

The variation within groups is defined as the within group variance or mean square within (MSW). To calculate this value we use a weighted sum of the variance for the individual groups. The weights are based on the number of observations in each group. This sum is divided by the number of degrees of freedom calculated by taking the total number of observations (N) and subtracting the number of groups (k).

$$MSW = \frac{\sum_{i=1}^{k} (n_i - 1)s_i^2}{N - k}$$

Table 5.9. Calculating means and variances

	Call center A	Call center B	Call center C	Call center D	4 Groups (k)
	136	124	142	149	
	145	131	145	157	
	139	128	139	154	
	132	130	145	155	
	141	129	143	151	
	143	135	141	156	
	138	132	138		
	139		146		
Count (n)	8	7	8	6	**Total count N = 29**
Mean ($\bar{x}_i$)	139.1	129.9	142.4	153.7	**Average of means $\bar{\bar{x}}$ = 141.3**
Variance (s_i^2)	16.4	11.8	8.6	9.5	

In this example:

$$MSW = \frac{(8-1) \times 16.4 + (7-1) \times 11.8 + (8-1) \times 8.6 + (6-1) \times 9.5}{(29-4)}$$

$$MSW = 11.73$$

Determine the Between Group Variation

Next, the between group variation or mean square between (*MSB*) is calculated. The mean square between is the variance between the group means. It is calculated using a weighted sum of the squared difference between the group mean ($\bar{x}_i$) and the average of the means ($\bar{\bar{x}}$). This sum is divided by the number of degrees of freedom. This is calculated by subtracting one from the number of groups (k). The following formula is used to calculate the mean square between (*MSB*):

$$MSB = \frac{\sum_{i=1}^{k} n_i (\bar{x}_i - \bar{\bar{x}})^2}{k-1}$$

Where n_i is the size of each group and $\bar{x}_i$ is the average for each group.

In this example,

MSB

$$= \frac{(8 \times (139.1 - 141.3)^2) + (7 \times (129.9 - 141.3)^2) + (8 \times (142.4 - 141.3)^2) + (6 \times (153.7 - 141.3)^2)}{4-1}$$

$$MSB = 626.89$$

Determine the F-Statistic

The F-statistic is the ratio of the mean square between (MSB) and the mean square within (MSW):

$$F = \frac{MSB}{MSW}$$

In this example:

$$F = \frac{626.89}{11.73}$$

$$F = 53.44$$

Test the Significance of the F-Statistic

Before we can test the significance of this value, we must determine the degrees of freedom (df) for the two mean squares (within and between).

The degrees of freedom for the mean square within (df_{within}) is calculated using the following formula:

$$df_{within} = N - k$$

where N is the total number of observations in all groups and k is the number of groups.

The degrees of freedom for the mean square between ($df_{between}$) is calculated using the following formula:

$$df_{between} = k - 1$$

where k is the number of groups.

In this example:

$$df_{between} = 4 - 1 = 3$$
$$df_{within} = 29 - 4 = 25$$

We already calculated the F-statistic to be 53.44. This number indicates that the mean variation between groups is much greater than the mean variation within groups due to errors. To test this, we look up the critical F-statistic from Appendix A.4. To find this critical value we need α (confidence level), v_1 ($df_{between}$), and v_2 (df_{within}). The critical value for the F-statistic is 3.01, as shown in Figure 5.18. Since the calculated F-statistic is greater than the critical value, we reject the null hypothesis. The means for the different call centers are not equal.

5.4 COMPARATIVE STATISTICS

5.4.1 Overview

Correlation analysis looks at associations between variables. For example, is there a relationship between interest rates and inflation or education level and income? The

α = 0.05

v_1 / v_2	1	2	3	4	5	6	7	8	9	10	12	15	20	24	30	40	60	120	∞
1	161.4	199.5	215.7	224.6	230.2	234.0	236.8	238.9	240.5	241.9	243.9	245.9	248.0	249.1	250.1	250.1	252.2	253.3	254.3
2	18.51	19.00	19.16	19.25	19.30	19.33	19.35	19.37	19.38	19.40	19.41	19.43	19.45	19.45	19.46	19.47	19.48	1949	19.50
3	10.13	9.55	9.28	9.12	9.01	8.94	8.89	8.85	8.81	8.79	8.74	8.70	8.66	8.64	8.62	8.59	8.57	8.55	8.53
4	7.71	6.94	6.59	6.39	6.26	6.16	6.09	6.04	6.00	5.96	5.91	5.86	5.80	5.77	5.75	5.72	5.69	5.66	5.63
5	6.61	5.79	5.41	5.19	5.05	4.95	4.88	4.82	4.77	4.74	4.68	4.62	4.56	4.53	4.50	4.46	4.43	4.40	4.36
6	5.99	5.14	4.76	4.53	4.39	4.28	4.21	4.15	4.10	4.06	4.00	3.94	3.87	3.84	3.81	3.77	3.74	3.70	3.67
7	5.59	4.74	4.35	4.12	3.97	3.87	3.79	3.73	3.68	3.64	3.57	3.51	3.44	3.41	3.38	3.34	3.30	3.27	3.23
8	5.32	4.46	4.07	3.84	3.69	3.58	3.50	3.44	3.39	3.35	3.28	3.22	3.15	3.12	3.08	3.04	3.01	2.97	2.93
9	5.12	4.26	3.86	3.63	3.48	3.37	3.29	3.23	3.18	3.14	3.07	3.01	2.94	2.90	2.86	2.83	2.79	2.75	2.71
10	4.96	4.10	3.71	3.48	3.33	3.22	3.14	3.07	3.02	2.98	2.91	2.85	2.77	2.74	2.70	2.66	2.62	2.58	2.54
11	4.84	3.98	3.59	3.36	3.20	3.09	3.01	2.95	2.90	2.85	2.79	2.72	2.65	2.61	2.57	2.53	2.49	2.45	2.40
12	4.75	3.89	3.49	3.26	3.11	3.00	2.91	2.85	2.80	2.75	2.69	2.62	2.54	2.51	2.47	2.43	2.38	2.34	2.30
13	4.67	3.81	3.41	3.18	3.03	2.92	2.83	2.77	2.71	2.67	2.60	2.53	2.46	2.42	2.38	2.34	2.30	2.25	2.21
14	4.60	3.74	3.34	3.11	2.96	2.85	2.76	2.70	2.65	2.60	2.53	2.46	2.39	2.35	2.31	2.27	2.22	2.18	2.13
15	4.54	3.68	3.29	3.06	2.90	2.79	2.71	2.64	2.59	2.54	2.48	2.40	2.33	2.29	2.25	2.20	2.16	2.11	2.07
16	4.49	3.63	3.24	3.01	2.85	2.74	2.66	2.59	2.54	2.49	2.42	2.35	2.28	2.24	2.19	2.15	2.11	2.06	2.01
17	4.45	3.59	3.20	2.96	2.81	2.70	2.61	2.55	2.49	2.45	2.38	2.31	2.23	2.19	2.15	2.10	2.06	2.01	1.96
18	4.41	3.55	3.16	2.93	2.77	2.66	2.58	2.51	2.46	2.41	2.34	2.27	2.19	2.16	2.11	2.06	2.02	1.97	1.92
19	4.38	3.52	3.13	2.90	2.74	2.63	2.54	2.48	2.42	2.38	2.31	2.23	2.16	2.11	2.07	2.03	1.98	1.93	1.88
20	4.35	3.49	3.10	2.87	2.71	2.60	2.51	2.45	2.39	2.35	2.28	2.20	2.12	2.08	2.04	1.99	1.95	1.90	1.84
21	4.32	3.47	3.07	2.84	2.68	2.57	2.49	2.42	2.37	2.32	2.25	2.18	2.10	2.05	2.01	1.96	1.92	1.87	1.81
22	4.30	3.44	3.05	2.82	2.66	2.55	2.46	2.40	2.34	2.30	2.23	2.15	2.07	2.03	1.98	1.94	1.89	1.84	1.78
23	4.28	3.42	3.03	2.80	2.64	2.53	2.44	2.37	2.32	2.27	2.20	2.13	2.05	2.01	1.96	1.91	1.86	1.84	1.76
24	4.26	3.40	3.01	2.78	2.62	2.51	2.42	2.36	2.30	2.25	2.18	2.11	2.03	1.98	1.94	1.89	1.84	1.79	1.73
25	4.24	3.39	2.99	2.76	2.60	2.49	2.40	2.34	2.28	2.24	2.16	2.09	2.01	1.96	1.92	1.87	1.82	1.77	1.71
26	4.23	3.37	2.98	2.74	2.59	2.47	2.39	2.32	2.27	2.22	2.15	2.07	1.99	1.95	1.90	1.85	1.80	1.75	1.69
27	4.21	3.35	2.96	2.73	2.57	2.46	2.37	2.31	2.25	2.20	2.13	2.06	1.97	1.93	1.88	1.84	1.79	1.73	1.67
28	4.20	3.34	2.95	2.71	2.56	2.45	2.36	2.29	2.24	2.19	2.12	2.04	1.96	1.91	1.87	1.82	1.77	1.71	1.65
29	4.18	3.33	2.93	2.70	2.55	2.43	2.35	2.28	2.22	2.18	2.10	2.03	1.94	1.90	1.85	1.81	1.75	1.70	1.64
32	4.17	3.32	2.92	2.69	2.53	2.42	2.33	2.27	2.21	2.16	2.09	2.01	1.93	1.89	1.84	1.79	1.74	1.68	1.62
40	4.08	3.23	2.84	2.61	2.45	2.34	2.25	2.18	2.12	2.08	2.00	1.92	1.84	1.79	1.74	1.69	1.64	1.58	1.51
60	4.00	3.15	2.76	2.53	2.37	2.25	2.17	2.10	2.04	1.99	1.92	1.84	1.75	1.70	1.65	1.59	1.53	1.47	1.39
120	3.92	3.07	2.68	2.45	2.29	2.17	2.09	2.02	1.96	1.91	1.83	1.75	1.66	1.61	1.55	1.50	1.43	1.35	1.25
∞	3.84	3.00	2.60	2.37	2.21	2.10	2.01	1.94	1.88	1.83	1.75	1.67	1.57	1.52	1.46	1.39	1.32	1.22	1.00

Critical F-statistic where α = 0.05, $df_{between} = 4$ (v_1) and $df_{within} = 24$ (v_2)

Adapted from E. S. Pearson and H. O. Hartley, *Biometrika Tables for Statisticians*, Vol. 1, 1958, pp. 157–63. Table 18, by permission of the Biometrika Trustees

Figure 5.18. Looking up critical F-statistic

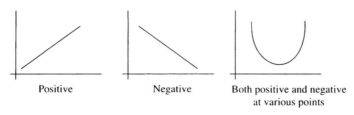

Figure 5.19. Relationships between two variables

existence of an association between variables does not imply that one variable causes another. Yet, understanding these relationships is useful for a number of reasons. For example, when building a predictive model, comparative statistics can help identify important variables to use.

The relationship between variables can be complex; however, a number of characteristics of the relationship can be measured:

- **Direction:** In comparing two variables, a *positive relationship* results when higher values in the first variable coincide with higher values in the second variable. In addition, lower values in the first variable coincide with lower values in the second variable. *Negative relationships* result when higher values in the first variable coincide with lower values in the second variable as well as lower values in the first variable coincide with higher values in the second variable. There are also situations where the relationship between the variables is more complex, having a combination of positive and negative relationships at various points. Figure 5.19 illustrates various scenarios for the relationship between variables.

- **Shape:** A relationship is linear when it is drawn as a straight line. As values for one variable change, the second variable changes proportionally. A non linear relationship is drawn as a curve indicating that as the first variable changes, the change in the second variable is not proportional. Figure 5.20 illustrates linear and non-linear relationships.

5.4.2 Visualizing Relationships

Where the data is categorical, the relationship between different values can be seen using a contingency table. For example, Table 5.10 illustrates the relationship

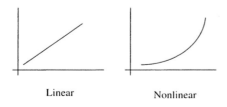

Linear   Nonlinear

Figure 5.20. Linear and nonlinear relationships

Table 5.10. Contingency table indicating results of a medical trial

	Takes medicine	Takes placebo	Total
Has strokes	213	342	555
No strokes	9,791	9,671	19,462
Totals	10,004	10,013	20,017

between whether a patient took a specific medicine and whether the patient had a stroke. Evaluating how these counts differ from the expected can be used to determine whether a relationship exists. The chi-square test, as previously described, can be used for this purpose.

A contingency table can also be used to crudely define the relationship between continuous variables. A table could be formed by converting the continuous variables into dichotomous variables through the setting of a cut off at the mean value. Values above the mean are assigned to one category and values below the mean are assigned to the other category.

It is usually more informative to explore the relationship between different continuous variables using a scatterplot. Figure 5.21 illustrates three scatterplots. In **a**, the relationship between the two variables is positive and from inspection appears to be linear. In **b**, there is a negative relationship between the variables and it also appears to be non linear. In **c**, it is difficult to see any relationship between the two variables.

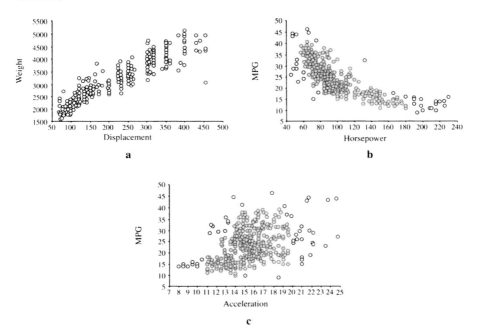

Figure 5.21. Illustrating different relationships using scatterplots

5.4.3 Correlation Coefficient (r)

For pairs of variables measured on an interval or ratio scale, a *correlation coefficient* (*r*) can be calculated. This value quantifies the linear relationship between the variables. It generates values ranging from -1.0 to $+1.0$. If an optimal straight line is drawn through the points on a scatterplot, then the value of *r* reflects how close to this line the points lie. Positive numbers indicate a positive correlation and negative numbers indicate a negative correlation. If *r* is around 0 then there appears to be little or no relationship between the variables.

For example, three scatterplots illustrate different values for *r* as shown in Figure 5.22. The first graph illustrates a good positive correlation, the second graph shows a negative correlation and the third graph illustrates a poor correlation.

The formula used to calculate *r* is shown here:

$$r = \frac{\sum_{i=1}^{n}(x_i - \bar{x})(y_i - \bar{y})}{(n-1)s_x s_y}$$

Two variables are considered in this formula: *x* and *y*. The individual values for *x* are x_i and the individual values for *y* are y_i. $\bar{x}$ is the mean of the *x* variable and $\bar{y}$ is the

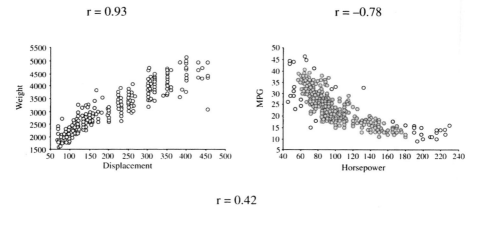

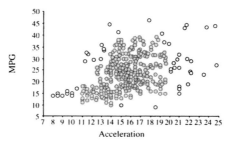

Figure 5.22.   Correlation coefficients for three relationships

Table 5.11.   Table of data with values for **x** and **y** variable

x	y
92	6.3
145	7.8
30	3
70	5.5
75	6.5
105	5.5
110	6.5
108	8
45	4
50	5
160	7.5
155	9
180	8.6
190	10
63	4.2
85	4.9
130	6
132	7

mean of the y variable. The number of observations is n. s_x is the standard deviations for x and s_y is the standard deviations for y.

To illustrate the calculation, two variables (x and y) are used and shown in Table 5.11. Plotting the two variables on a scatterplot indicates there is a positive correlation between these two variables, as shown in Figure 5.23. The specific value

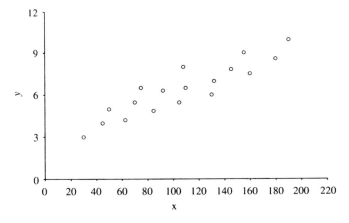

Figure 5.23.   Scatterplot showing relationship between x and y variables

Table 5.12. Table showing the calculation of the correlation coefficient

x_i	y_i	$(x_i - \bar{x})$	$(y_i - \bar{y})$	$(x_i - \bar{x})(y_i - \bar{y})$
92	6.3	−14.94	−0.11	1.58
145	7.8	38.06	1.39	53.07
30	3	−76.94	−3.41	262.04
70	5.5	−36.94	−0.91	33.46
75	6.5	−31.94	0.09	−3.02
105	5.5	−1.94	−0.91	1.76
110	6.5	3.06	0.094	0.29
108	8	1.06	1.59	1.68
45	4	−61.94	−2.41	149.01
50	5	−56.94	−1.41	80.04
160	7.5	53.06	1.09	58.07
155	9	48.06	2.59	124.68
180	8.6	73.06	2.19	160.32
190	10	83.06	3.59	298.54
63	4.2	−43.94	−2.21	96.92
85	4.9	−21.94	−1.51	33.04
130	6	23.06	−0.41	−9.35
132	7	25.06	0.59	14.89
$\bar{x} = 106.94$	$\bar{y} = 6.41$			Sum = 1357.01
$s_x = 47.28$	$s_y = 1.86$			

of r is calculated using Table 5.12:

$$r = \frac{\sum_{i=1}^{n}(x_i - \bar{x})(y_i - \bar{y})}{(n-1)s_x s_y}$$

$$r = \frac{1357.01}{(18-1)(47.28)(1.86)}$$

$$r = 0.91$$

5.4.4 Correlation Analysis for More Than Two Variables

When exploring data, it is useful to visualize the relationships between all variables in a data set. A matrix representation can be a useful presentation of this information. In this example, five variables relating to a data set of cars are presented: **Displacement, Horsepower, Weight, Acceleration, MPG**. The relationship (r) between each pair of variables is shown in Table 5.13. The correlation analysis for these variables can also be plotted using a matrix of scatterplots, as shown in Figure 5.24.

Table 5.13.   Table displaying values for the correlation coefficient for five variables

	Displacement	Horsepower	Weight	Acceleration	MPG
Displacement	1	0.9	0.93	−0.54	−0.81
Horsepower	0.9	1	0.86	−0.69	−0.78
Weight	0.93	0.86	1	−0.42	−0.83
Acceleration	−0.54	−0.69	−0.42	1	0.42
MPG	−0.81	−0.78	−0.83	0.42	1

The correlation coefficient is often squared (r^2) to represent the percentage of the variation that is explained by the regression line. For example, Table 5.14 illustrates the calculation for r^2 for the five variables illustrated in the scatterplot matrix (Figure 5.24).

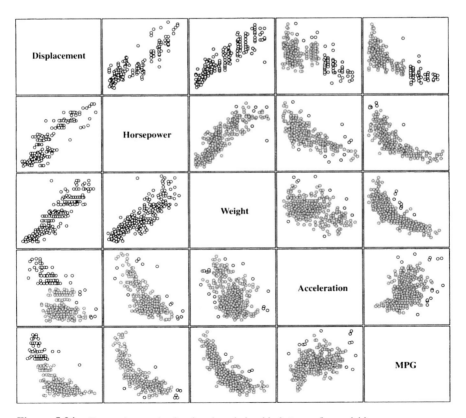

Figure 5.24.   Scatterplot matrix showing the relationship between five variables

Table 5.14.   Table displaying the value for r^2 for five variables

	Displacement	Horsepower	Weight	Acceleration	MPG
Displacement	1	0.81	0.87	0.29	0.66
Horsepower	0.81	1	0.74	0.48	0.61
Weight	0.87	0.74	1	0.18	0.69
Acceleration	0.29	0.48	0.18	1	0.18
MPG	0.66	0.61	0.69	0.18	1

5.5 SUMMARY

Central Tendency

Mode: Most common value

Median: Middle value

$$\text{Mean}: \bar{x} = \frac{\sum_{i=1}^{n} x_i}{n}$$

Variation

Range: high–low

Quartiles: Q1 (25%), Q2 (50%), Q3 (75%)

$$\text{Variance}: s^2 = \frac{\sum_{i=1}^{n} (x_i - \bar{x})^2}{n - 1}$$

$$\text{Standard deviation}: s = \sqrt{\frac{\sum_{i=1}^{n} (x_i - \bar{x})^2}{n - 1}}$$

$$z\text{-score}: z = \frac{x_i - \bar{x}}{s}$$

Confidence Levels

$$\text{Mean} (> 30 \text{ observations}) : \bar{x} \pm z_C \frac{s}{\sqrt{n}}$$

$$\text{Mean} (< 30 \text{ observations}) : \bar{x} \pm t_C \frac{s}{\sqrt{n}}$$

$$\text{Proportion}: p \pm z_C \sqrt{\frac{p(1 - p)}{n}}$$

Hypothesis Test

Specify null (e.g. H_0: $\mu = \mu_0$) and alternative hypothesis (e.g. H_a: $\mu > \mu_0$)

Select significance level (e.g. $\alpha = 0.05$)

Compute test statistics ($t-$ or $z-$)

Determine critical value for t or z using $\alpha/2$ for two sided tests

Reject the null hypothesis if test statistic fall in the "reject H_0" region

Comparing Groups

When comparing more than two groups, use:

Chi-square test for categorical data

One-way analysis of variance test for continuous.

Comparing Variables

Correlation coefficient (r): $\quad r = \dfrac{\displaystyle\sum_{i=1}^{n}(x_i - \bar{x})(y_i - \bar{y})}{(n-1)s_x s_y}$

5.6 EXERCISES

Table 5.15 presents the ages for a number of individuals.

1. Calculate the following statistics for the variable **Age**:
   a. Mode
   b. Median
   c. Mean

Table 5.15. Table with variables **Name** and **Age**

Name	Age
P.Lee	35
R.Jones	52
J.Smith	45
A.Patel	70
M.Owen	24
S.Green	43
N.Cook	68
W.Hands	77
P.Rice	45
F.Marsh	28

    d. Range

    e. Variance

    f. Standard deviation

    g. *z-score*

    h. Skewness

    i. Kurtosis

2. An insurance company wanted to understand the time to process an insurance claim. They timed a random sample of 47 claims and determined that it took on average 25 minutes per claim and the standard deviation was calculated to be 3. With a confidence level of 95%, what is the confidence interval?

3. An electronics company wishes to understand, for all customers that purchased a computer, how many will buy a printer at the same time. To test this, the company interviews a random sample of 300 customers and it was determined that 138 bought a printer. With a confidence level of 99%, what is the confidence interval for the proportion of customers buying a printer at the same time as a computer?

4. A phone company wishes to make a claim that the average connection time in the US is less than two seconds (i.e. the time after you dial a number before the call starts to ring). To test this, the company measures 50 randomly selected calls and the average time was 1.9 seconds with a standard deviation of 0.26. Using this information and a 95% confidence level:

    a. Specify the null and alternative hypothesis

    b. Calculate the hypothesis score

    c. Calculate a *p-value*

    d. Determine whether the phone company can make the claim

5. A bank wishes to make a claim that more than 90% of their customers are pleased with the level of service they receive. To test this claim, a random sample of 100 customers were questioned and 91 answered that they were pleased with the service. The bank wishes to make the claim at a 95% confidence level. Using this information:

    a. Specify the null and alternative hypothesis

    b. Calculate the hypothesis score

    c. Calculate a *p-value*

    d. Determine whether the bank can make the claim

6. A company that produces tomato plant fertilizer wishes to make a claim that their fertilizer (X) results in taller tomato plants than a competitor product (Y). Under highly controlled conditions, 50 plants were grown using X and 50 plants grown using Y and the height of the plants were measured. The average height of the plants grown with fertilizer X is 0.36 meters with a standard deviation of 0.035. The average height of the plants grown with fertilizer Y was 0.34 with a standard deviation of 0.036. Using a 95% confidence limit:

    a. Specify the null and alternative hypothesis

    b. Calculate the hypothesis score

    c. Calculate a *p-value*

    d. Determine whether the company can make the claim

Table 5.16. Contingency table showing defective products produced using material from two manufacturers

	Defective	Not defective	
Manufacturer A	7	98	105
Manufacturer B	5	97	102
Totals	12	195	207

7. A producer of kettles wishes to assess whether a new supplier of steel (B) results in kettles with fewer defects than the existing supplier (A). To test this, the company collects a number of kettles generated from both suppliers to examine the kettles for defects. Table 5.16 summarizes the counts. Using a 95% confidence limit:
   a. Specify the null and alternative hypothesis
   b. Calculate the hypothesis score
   c. Calculate a *p-value*
   d. Determine whether the company can make the claim

8. A construction company wants to understand whether there is a difference in wear for different types of gloves (P and Q). 40 employees wear P gloves on one hand and Q gloves on the other. The hands are randomized. The wear of the gloves were recorded and the average difference calculated. The average difference was 0.34 with a standard deviation of 0.14. Using a 95% confidence limit:
   a. Specify the null and alternative hypothesis
   b. Calculate the hypothesis score
   c. Calculate a *p-value*
   d. Determine whether the company can make the claim

9. A producer of magnets wishes to understand whether there is a difference between four suppliers (A, B, C, and D) of alloys used in the production of the magnets. Magnets from the four suppliers are randomly selected and the magnets are recorded as either satisfactory or not satisfactory as shown in Table 5.17. With a 95% confidence limit and using this information:
   a. Specify the null and alternative hypothesis
   b. Calculate chi-square
   c. Determine whether the company can make the claim

Table 5.17. Contingency table showing product satisfaction using materials from four suppliers

	Satisfactory	Not satisfactory	Total
Supplier A	28	2	30
Supplier B	27	3	30
Supplier C	29	1	30
Supplier D	26	4	30
Total	110	10	120

Table 5.18.   Table of snacks per packet produced by four machines

Machine 1	Machine 2	Machine 3	Machine 4
50	51	49	52
51	52	51	50
50	50	50	53
52	51	51	51
50	53	49	50
49	50	51	50
52	51	49	49
49	50	49	51

Table 5.19.   Table showing observations for variables **Amount of Sun** and **Tree Height**

Amount of Sun	Tree Height
2.4	3
2.6	3.1
2.9	3.1
3.4	3.5
3.8	3.7
4.2	3.8
4.5	4.1
5.1	4.3
5.8	5.1

10. A food producer creates packets of snacks using four machines (1, 2, 3, 4). The number of snacks per packet is recorded for a random collection of samples from the four machines, as shown in Table 5.18. The company wishes to know if there is a difference between the four machines. Using a 95% confidence limit:

   a. Specify the null and alternative hypothesis

   b. Calculate the F-statistic

   c. Determine whether the company can make the claim

11. In a highly controlled experiment, a biologist was investigating whether there exists a relationship between the height of a tree and their exposure to the sun. The biologist recorded the results in Table 5.19. Calculate the correlation coefficient between these two columns.

5.7   FURTHER READING

This chapter has focused on techniques for summarizing, making statements about population from samples, and quantifying relationships in the data. There are a number of introductory statistical books that provide an overview of the theory behind these techniques including the central limits theorem: Donnelly (2004), Freedman (1997), Rumsey (2003), Kachigan (1991), and Levine (2005).

The following web sites contain information on statistics and other data analysis methods:

http://www.statsoft.com/textbook/stathome.html

http://www.itl.nist.gov/div898/handbook/index.htm

The following web site contains information on the R-Project on statistical computing:

http://www.r-project.org/

Chapter 6

Grouping

6.1 INTRODUCTION

6.1.1 Overview

Dividing a data set into smaller subsets of related observations or groups is important for exploratory data analysis and data mining for a number of reasons:

- **Finding hidden relationships:** Grouping methods organize observations in different ways. Looking at the data from these different angles will allow us to find relationships that are not obvious from a summary alone. For example, a data set of retail transactions is grouped and these groups are used to find nontrivial associations, such as customers who purchase doormats often purchase umbrellas at the same time.

- **Becoming familiar with the data:** Before using a data set to create a predictive model, it is beneficial to become highly familiar with the contents of the set. Grouping methods allows us to discover which types of observations are present in the data. In the following example, a database of medical records will be used to create a general model for predicting a number of medical conditions. Before creating the model, the data set is characterized by grouping the observations. This reveals that a significant portion of the data consists of young female patients having flu. It would appear that the data set is not evenly stratified across the model target population, that is, both male and female patients with a variety of conditions. Therefore, it may be necessary to create from these observations a *diverse* subset that matches more closely the target population.

- **Segmentation:** Techniques for grouping data may lead to divisions that simplify the data for analysis. For example, when building a model that predicts car fuel efficiency, it may be possible to group the data to reflect the underlying technology platforms the cars were built on. Generating a model for each of these 'platform-based' subsets will result in simpler models.

6.1.2   Grouping by Values or Ranges

One way of creating a group is to search or query the data set. Each query would bring back a subset of observations. This set could then be examined to determine whether some interesting relationship exists. For example, in looking for hidden relationships that influence car fuel efficiency, we may query the data set in a variety of ways. The query could be by a single value, such as where the number of cylinders is four. Alternatively, a range of values could be used, such as all cars with **Weight** less than 4000. Boolean combinations of query terms could also be used to create more complex queries, for example cars where **Cylinders** is equal to six and **Weight** is greater than 5000. The following illustrates two queries:

Query 1: All cars where **Horsepower** is greater than or equal to 160 AND **Weight** is greater than or equal to 4000.

This query will bring back all observations where **Horsepower** is greater than or equal to 160 and **Weight** is greater than or equal to 4000. A sample extracted from the 31 observations returned is shown in Table 6.1. The relationship of the 31 observations to car fuel efficiency can be seen in Figure 6.1, with the 31 observations highlighted. Cars containing the values in the query (i.e. heavy vehicles with high horsepower) seem to be associated with low fuel-efficient vehicles.

Query 2: All cars where **Horsepower** is less than 80 AND **Weight** is less than 2500.

Table 6.1.   Cars where **Horsepower** $\geq$ 160 and **Weight** $\geq$ 4000

Names	Cylinders	Displace-ment	Horse-power	Weight	Accele-ration	Model/Year	Origin	MPG
Ford Galaxie 500	8	429	198	4,341	10	1970	1	15
Chevrolet Impala	8	454	220	4,354	9	1970	1	14
Plymouth Fury III	8	440	215	4,312	8.5	1970	1	14
Pontiac Catalina	8	455	225	4,425	10	1970	1	14
Ford F250	8	360	215	4,615	14	1970	1	10
Chevy C20	8	307	200	4,376	15	1970	1	10
Dodge D200	8	318	210	4,382	13.5	1970	1	11
Hi 1200d	8	304	193	4,732	18.5	1970	1	9
Pontiac Catalina Brougham	8	400	175	4,464	11.5	1971	1	14
Dodge Monaco (SW)	8	383	180	4,955	11.5	1971	1	12

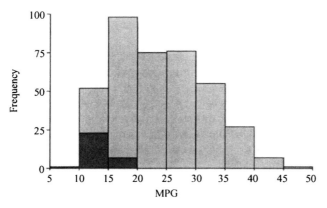

Figure 6.1. Highlighted observations where **Horsepower** ≥ 160 and **Weight** ≥ 4000

A sample extracted from the 97 observations returned is shown in Table 6.2. The relationship of the 97 observations to car fuel efficiency can be seen in Figure 6.2, with the 97 observations highlighted. Cars containing the combinations of values in the query (i.e. light vehicles with low horsepower) seem to be associated with good fuel efficiency.

By grouping the data in different ways and looking to see how the groups influence car fuel efficiency (**MPG**) we can start to uncover hidden relationships. In addition, we could assess these claims using hypothesis tests described in Section 5.2.3. Unfortunately, an exhaustive exercise of this nature would not be feasible. Fortunately, many computational methods will group observations efficiently by values or ranges without resorting to an exhaustive search for all combinations of values.

6.1.3 Similarity Measures

Any method of grouping needs to have an understanding for how similar observations are to each other. One method, as described in the previous section, is to define groups sharing the same values or ranges of values. An alternative method is to

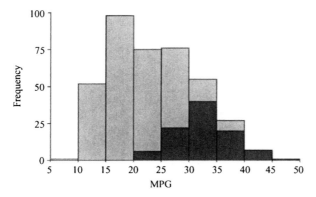

Figure 6.2. Highlighted observations where **Horsepower** < 80 and **Weight** < 2500

Table 6.2.   Table of cars where **Horsepower** < 80 and **Weight** < 2500

Names	Cylinders	Displace ment	Horse- power	Weight	Accele- ration	Model/ Year	Origin	MPG
Volkswagen 1131 Deluxe Sedan	4	97	46	1,835	20.5	1970	2	26
Chevrolet Vega (SW)	4	140	72	2,408	19	1971	1	22
Peugeot 304	4	79	70	2,074	19.5	1971	2	30
Fiat 124B	4	88	76	2,065	14.5	1971	2	30
Toyota Corolla 1200	4	71	65	1,773	19	1971	3	31
Datsun 1200	4	72	69	1,613	18	1971	3	35
Volkswagen model 111	4	97	60	1,834	19	1971	2	27
Plymouth Cricket	4	91	70	1,955	20.5	1971	1	26
Volkswagen type 3	4	97	54	2,254	23.5	1972	2	23
Renault 12 (SW)	4	96	69	2,189	18	1972	2	26

determine whether observations are more generally similar. To determine how similar two observations are to each other we need to compute the *distance* between them. To illustrate the concept of distance we will use a simple example with two observations and two variables (Table 6.3). The physical distance between the two observations can be seen by plotting them on a scatterplot (Figure 6.3). In this example, the distance between the two observations is calculated using simple trigonometry:

$$x = 7 - 2 = 5$$
$$y = 8 - 3 = 5$$
$$d = \sqrt{x^2 + y^2} = \sqrt{25 + 25} = 7.07$$

We can extend this concept of distance between observations with more than two variables. This calculation is called the *Euclidean* distance (d) and the formula is shown:

$$d = \sqrt{\sum_{i=1}^{n} (p_i - q_i)^2}$$

Table 6.3.   Table showing two observations (A and B)

	Variable 1	Variable 2
A	2	3
B	7	8

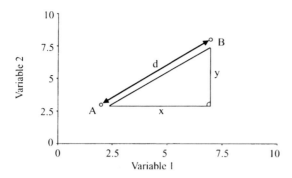

Figure 6.3. Distance between two observations (A and B)

It calculates the distance between two observations p and q, where each observation has n variables. To illustrate the Euclidean distance calculation for observations with more than two variables, we will use Table 6.4.

The Euclidean distance between A and B is

$$d_{A-B} = \sqrt{(0.7 - 0.6)^2 + (0.8 - 0.8)^2 + (0.4 - 0.5)^2 + (0.5 - 0.4)^2 + (0.2 - 0.2)^2}$$
$$d_{A-B} = 0.17$$

The Euclidean distance between A and C is

$$d_{A-C} = \sqrt{(0.7 - 0.8)^2 + (0.8 - 0.9)^2 + (0.4 - 0.7)^2 + (0.5 - 0.8)^2 + (0.2 - 0.9)^2}$$
$$d_{A-C} = 0.83$$

The Euclidean distance between B and C is

$$d_{B-C} = \sqrt{(0.6 - 0.8)^2 + (0.8 - 0.9)^2 + (0.5 - 0.7)^2 + (0.4 - 0.8)^2 + (0.2 - 0.9)^2}$$
$$d_{B-C} = 0.86$$

The distance between A and B is 0.17, indicating that there is more similarity between these observations than A and C (0.83). C is not so closely related to either A or B. This can be seen in Figure 6.4 where the values for each variable are plotted.

Table 6.4. Three observations with values for five variables

Name	Variable 1	Variable 2	Variable 3	Variable 4	Variable 5
A	0.7	0.8	0.4	0.5	0.2
B	0.6	0.8	0.5	0.4	0.2
C	0.8	0.9	0.7	0.8	0.9

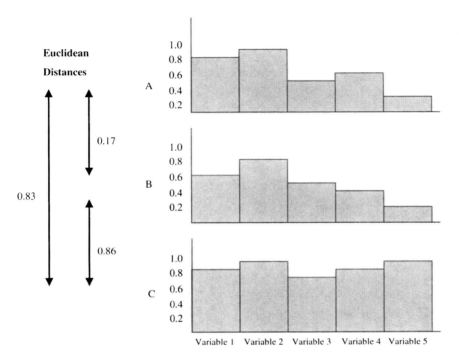

Figure 6.4. Distances between observations with five variables

The shape of histograms A and B are similar, whereas the shape of histogram C is not similar to A or B.

The Euclidean distance handles continuous variables. Another method that handles only binary variables is the *Jaccard* distance. The contingency table shown in Table 6.5 is used to calculate the Jaccard distance between two observations that have been measured over a series of binary variables.

The table shows the following counts:

- $Count_{11}$: Count of all variables that are 1 in "Observation 1" and 1 in "Observation 2".

- $Count_{10}$: Count of all variables that are 1 in "Observation 1" and 0 in "Observation 2".

Table 6.5. Table showing the relationship between two observations measured using a series of binary variables

		Observation 2	
		1	**0**
Observation 1	**1**	$Count_{11}$	$Count_{10}$
	0	$Count_{01}$	$Count_{00}$

Table 6.6.   Table of observations with values for five binary variables

Name	Variable 1	Variable 2	Variable 3	Variable 4	Variable 5
A	1	1	0	0	1
B	1	1	0	0	0
C	0	0	1	1	1

- $Count_{01}$: Count of all variables that are 0 in "Observation 1" and 1 in "Observation 2".
- $Count_{00}$: Count of all variables that are 0 in "Observation 1" and 0 in "Observation 2".

The following formula is used to calculate the Jaccard distance:

$$d = \frac{Count_{10} + Count_{01}}{Count_{11} + Count_{10} + Count_{01}}$$

The Jaccard distance is illustrated using Table 6.6.

The Jaccard distance between A and B is:

$$d_{A-B} = (1 + 0)/(2 + 1 + 0) = 0.33$$

The Jaccard distance between A and C is:

$$d_{A-C} = (2 + 2)/(1 + 2 + 2) = 0.8$$

The Jaccard distance between B and C is:

$$d_{B-C} = (2 + 3)/(0 + 2 + 3) = 1.0$$

The Euclidean and Jaccard distance measures are two examples for determining the distance between observations. Other techniques include Mahalanobis, City Block, Minkowski, Cosine, Spearman, Hamming and Chebuchev (see the further reading section for references on these methods).

6.1.4   Grouping Approaches

There exist numerous automatic methods for grouping observations. These techniques are commonly used in a variety of exploratory data analysis and data mining situations. When selecting a grouping method, there are a number of issues (in addition to defining how similar two or more observations are to each other) to consider:

- **Supervised versus unsupervised:** One distinction between the different methods is whether they use the response variable to guide how the groups are generated. Methods that do not use any data to guide how the groups are generated are called *unsupervised* methods, whereas methods that make use of the response variable to guide group generation are called *supervised* methods. For example, a data set of cars could be grouped using an unsupervised method. The groups generated would be based on general classes of cars. Alternatively, we could group the cars using car fuel efficiency to direct the grouping. This would generate groups directed towards finding hidden relationships between groups of cars and car fuel efficiency. Now, if we were to repeat the exercise using a different goal, for example, car acceleration, the data would be grouped differently. In this situation the groups are directed towards finding hidden relationships associated with car acceleration.

- **Type of variables:** Certain grouping methods will only accept categorical data, whereas others only accept continuous data. Other techniques handle all types of data. Understanding these limitations will allow you to select the appropriate method. Alternatively, you could decide to restrict the variables used in the method or perform a transformation on the data.

- **Data set size limit:** There are methods that only work with data sets less than a certain size. Others work best with data sets over a certain size. Understanding the limitations placed on the number of observations and/or number of variables helps in the selection of particular methods. In situations where the data set is too large to process, one solution would be to segment the data prior to grouping.

- **Interpretable and actionable:** Certain grouping methods generate results that are easy to interpret, whereas other methods require additional analysis to interpret the results. How the grouping results will be used influences which grouping methods should be selected.

- **Overlapping groups:** In certain grouping methods, observations can only fall in one group. There are other grouping methods where the same observation may be a member of multiple groups.

A related topic to grouping is the identification of outliers, that is, observations that do not look like anything else. Single or small numbers of observations that fall into groups on their own are considered outliers. A data set where most of the observations fall into separate groups would be described as diverse. To create a diverse subset, representative observations from each group may be selected. Other methods for assessing outliers are discussed at the end of the chapter.

This chapter describes three popular methods for grouping data sets: clustering, associative rules, and decision trees. They cover different criteria for generating groups as well as supervised and unsupervised approaches. All approaches have advantages and disadvantages, and all provide different insights into the data. It is often informative to combine these grouping methods with other data analysis/data mining techniques, such as hypothesis tests to evaluate any claims made concerning the groups. The different methods have parameters that can be modified to optimize the results and these are described.

6.2 CLUSTERING

6.2.1 Overview

Clustering will group the data into sets of related observations or clusters. Observations within each group are more similar to other observations within the group than to observations within any other group. Clustering is an unsupervised method for grouping. To illustrate the process of clustering, a set of observations are shown on the scatterplot in Figure 6.5. These observations are plotted using two hypothetical dimensions and the similarity between the observations is proportional to the physical distance between the observations. There are two clear regions that could be considered as clusters: Cluster A and Cluster B. Clustering is a flexible approach to grouping. For example, based on the criteria for clustering the observations, observation X was not judged to be a member of Cluster A. However, if a different criterion was used, X may have been included in Cluster A. Clustering not only assists in identifying groups of related observations, it also locates observations that are not similar to others, that is outliers, since they fall into groups of their own.

Clustering has the following advantages:

- **Flexible:** There are many ways of adjusting how clustering is implemented, including options for determining the similarity between two observations and options for selecting the size of the clusters.

- **Hierarchical and nonhierarchical approaches:** Certain clustering techniques organize the data sets hierarchically, which may provide additional insight into the problem under investigation. For example, when clustering a genomic data set, hierarchical clustering may provide insight into evolutionary processes that have taken place since genes mutate over time. Other methods only generate lists of clusters based on a pre-defined number.

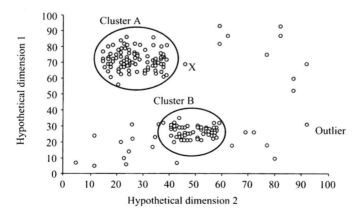

Figure 6.5. Illustration of clusters and outliers

Clustering has the following limitations:

- **Subjective:** Different problems will require different clustering options and specifying these options requires repeatedly examining the results and adjusting the clustering options accordingly.

- **Interpretation:** Observations are grouped together based on some measure of similarity. Making sense of a particular cluster may require additional analysis in order to take some action based on the results of a grouping.

- **Speed:** There are many techniques for clustering data and it can be time-consuming to generate the clusters, especially for large data sets.

- **Size limitations:** Certain techniques for clustering have limitations on the number of observations that they can process.

Two clustering techniques will be described: hierarchical agglomerative clustering and k-means clustering. Additional clustering methods will be described in the further readings section of this chapter.

6.2.2 Hierarchical Agglomerative Clustering

Overview

Hierarchical agglomerative clustering is an example of a hierarchical method for grouping observations. It uses a "bottom-up" approach to clustering as it starts with each observation as a member of a separate cluster and progressively merges clusters together until all observations are a member of a final single cluster. The major limitation of hierarchical agglomerative clustering is that it is normally limited to small data sets (often less than 10,000 observations) and the speed to generate the hierarchical tree can be slow for higher numbers of observations.

To illustrate the process of hierarchical agglomerative clustering, we will use the data set shown in Table 6.7 containing 14 observations, each measured over five variables. In this case the variables are all measured on the same scale; however, where variables are measured on different scales they should be normalized to a comparable range (e.g. 0 to 1). This is to avoid any one or more variables having a disproportionate weight and creating a bias in the analysis.

First, the distance between all combinations of observations is calculated. The method for assessing the distance along with which variables to include in the calculation should be set prior to clustering. The two closest observations are identified and are merged into a single cluster. These two observations from now on will be considered a single group. Next, all observations (minus the two that have been merged into a cluster) along with the newly created cluster are compared to see which observation or cluster should be joined into the next cluster. We are now analyzing both individual observations and clusters. The distance between a single observation and a cluster is determined based on a pre-defined linkage rule. The different types of linkage rules will be described in the next section. All

Table 6.7. Table of observations to cluster

Name	Variable 1	Variable 2	Variable 3	Variable 4	Variable 5
A	7	8	4	5	2
B	6	8	5	4	2
C	8	9	7	8	9
D	6	7	7	7	8
E	1	2	5	3	4
F	3	4	5	3	5
G	7	8	8	6	6
H	8	9	6	5	5
I	2	3	5	6	5
J	1	2	4	4	2
K	3	2	6	5	7
L	2	5	6	8	9
M	3	5	4	6	3
N	3	5	5	6	3

distances between all combinations of groups and observations are considered and the smallest distance is selected. The process continues until there are no more clusters to join.

Figure 6.6 illustrates the process. In step 1, it is determined that observations M and N are the closest and they are linked into a cluster, as shown. The horizontal

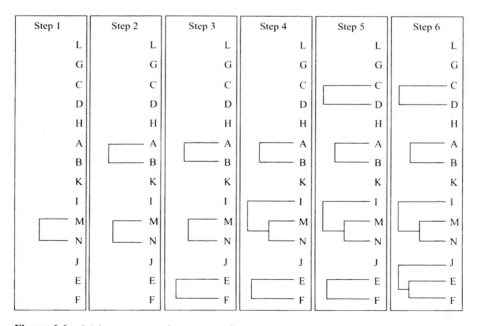

Figure 6.6. Joining process used to generate clusters

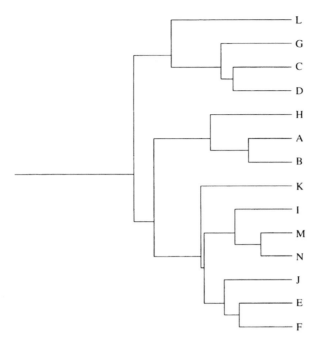

Figure 6.7. Complete hierarchical clustering of 14 observations

length of the lines joining M and N reflects the distance at which the cluster was formed. From now on M and N will not be considered individually, but only as a cluster. In step 2, distances between all observations (except M and N), as well as the cluster containing M and N, are calculated. To determine the distance between the individual observations and the cluster containing M and N, the average linkage rule was used (described in the next section). It is now determined that A and B should be joined as shown. Once again, all distances between the remaining ungrouped observations and the newly created clusters are calculated, and the smallest distance selected. In step 4, the shortest distance is between observation I and the cluster containing M and N. This process continues until only one cluster remains which contains all the observations. Figure 6.7 shows the complete hierarchical clustering for all 14 observations.

Linkage Rules

A linkage rule is used to determine a distance between an observation (or a group) and an already identified group. In Figure 6.8, two clusters have already been identified: Cluster A and Cluster B. We now wish to determine whether observation X is a member of cluster A.

There are many ways for determining the distance between an observation and an already established cluster and include average linkage, single linkage, and complete linkage. These alternatives are illustrated in Figure 6.9.

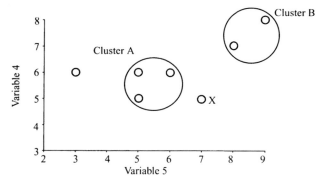

Figure 6.8.   Determining whether an observation X belongs to Cluster A

- **Average linkage:** the distance between all members of the cluster (e.g. a, b, and c) and the observation under consideration (e.g. X) are determined and the average is calculated.

- **Single linkage:** the distance between all members of the cluster (e.g. a, b, and c) and the observation under consideration (e.g. X) are determined and the smallest is selected.

- **Complete linkage:** the distance between all members of the cluster (e.g. a, b, and c) and the observation under consideration (e.g. X) are determined and the highest is selected.

These different linkage rules change how the final hierarchical clustering is presented. Figure 6.10 shows the hierarchical clustering of the same set of observations using the average linkage, single linkage, and complete linkage rules.

Creating Clusters

Up to this point, a tree has been generated showing the similarity between observations and clusters. To divide a data set into a series of clusters from this tree, we must determine a distance at which the clusters are to be created. Where this distance intersects with a line on the tree, a cluster is formed. Figure 6.11 illustrates this point. A distance is selected, as shown by the vertical line. Where this vertical

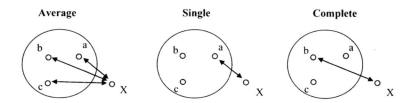

Figure 6.9.   Different linkage rules

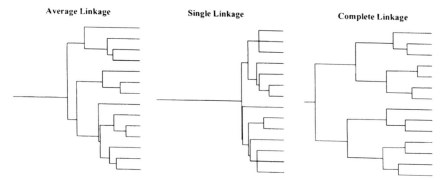

Figure 6.10.   Clustering using different linkage rules

line intersects with the tree (shown by the circles), four clusters are selected. Cluster 1 contains a single observation (L) and at this distance, it would be considered an outlier. Cluster 2 (G, C, D) and Cluster 3 (H, A, B) each contain three observations and the largest group is Cluster 4 (K, I, M, N, J, E, F) with seven observations. Observations will only be present in a single cluster.

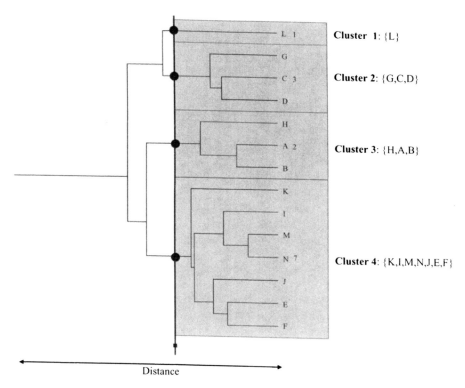

Figure 6.11.   Generating four clusters by specifying a distance

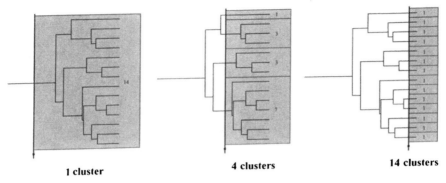

1 cluster   **4 clusters**   **14 clusters**

Figure 6.12. Adjusting the distance to generate different numbers of clusters

Adjusting the cut-off distance will change the number of clusters created. Figure 6.12 shows the selection of a single cluster when the distance cutoff is at the left. When the distance cutoff is placed to the far right, each observation will be in its own cluster. A cutoff placed between these two extremes will result in groups of various sizes. Cutoffs towards the left will result in fewer clusters with more diverse observations within each cluster. Cutoffs towards the right will result in more clusters with more similar observations within each cluster.

Example

The following example uses a data set of 392 cars that will be explored using hierarchical agglomerative clustering. A portion of the data table is shown in Table 6.8.

Table 6.8. Table of car observations

Names	Cylinders	Displace-ment	Horse power	Weight	Accele-ration	Model/ Year	Origin	MPG
Chevrolet Chevelle Malibu	8	307	130	3,504	12	1970	1	18
Buick Skylark 320	8	350	165	3,693	11.5	1970	1	15
Plymouth Satellite	8	318	150	3,436	11	1970	1	18
Amc Rebel SST	8	304	150	3,433	12	1970	1	16
Ford Torino	8	302	140	3,449	10.5	1970	1	17
Ford Galaxie 500	8	429	198	4,341	10	1970	1	15
Chevrolet Impala	8	454	220	4,354	9	1970	1	14
Plymouth Fury III	8	440	215	4,312	8.5	1970	1	14

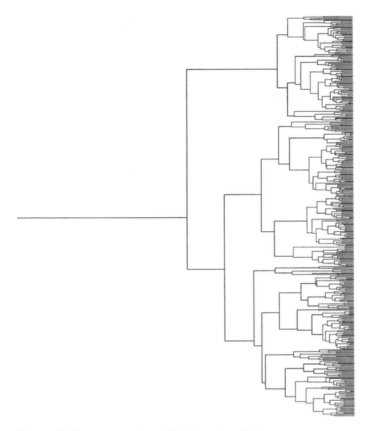

Figure 6.13. Complete hierarchical clustering of 392 cars

The data set was clustered using the Euclidean distance and the average linkage joining rule. The following variables were used in the clustering: **Cylinders, Displacement, Horsepower, Weight, Acceleration, Model Year** and **Origin. MPG** (miles per gallon) was not used in the clustering but will be considered later. Figure 6.13 shows the hierarchical tree generated.

The process of generating the tree is typically the most time consuming part of the process. Once the tree has been generated, it is usually possible to interactively explore the clusters. For example, in Figure 6.14 a distance cutoff has been set; such that, the data is divided into three clusters.

- **Cluster 1:** A cluster containing 103 observations is selected and shown in Figure 6.15. In addition to showing the tree, a table of charts illustrates the composition of the cluster. The highlighted histogram region corresponds to the distribution of Cluster 1 observations. The darker **MPG** box plot corresponds to the 103 selected observations with the lower and lighter box plot corresponding to all the observations. The cluster comprises of vehicles with eight **cylinders**, high **displacement**, **horsepower**, and **weight**,

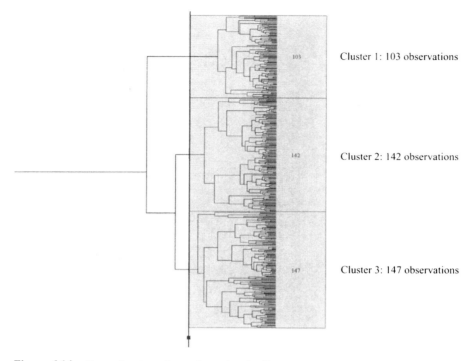

Figure 6.14. Generating three clusters by setting the distance

but with low **acceleration**. The majority was made in the 1970s and all were made in the US (**origin** 1 is the US). **MPG** was not used in the clustering process; however, it can be seen from the histogram and box plot that these vehicles have some of the lowest fuel efficiency.

- **Cluster 2:** A cluster of 142 observations is shown in Figure 6.16. The group comprises of vehicles with four or six **cylinders**, moderate-to-low **displacement** and **horsepower** with low **weight** and **acceleration**. They were all made in the US (**origin** 1 is the US) throughout the 1970s and 1980s. It can be seen that the fuel efficiency is similar to the average fuel efficiency for all cars.

- **Cluster 3:** A cluster of 147 observations is shown in Figure 6.17. The group is primarily made of vehicles with four **cylinders**, with low **displacement**, **horsepower**, **weight**, and average **acceleration**. They were all made outside the US (**origin** 2 is Europe and **origin** 3 is Japan) throughout the 1970s and 1980s. It can be seen that the fuel efficiency is higher for these vehicles than the average fuel efficiency for all cars reported.

To explore the data set further, we can adjust the distance cutoff to generate different numbers of clusters. In this case, the distance was set to create 16 clusters, as shown in Figure 6.18.

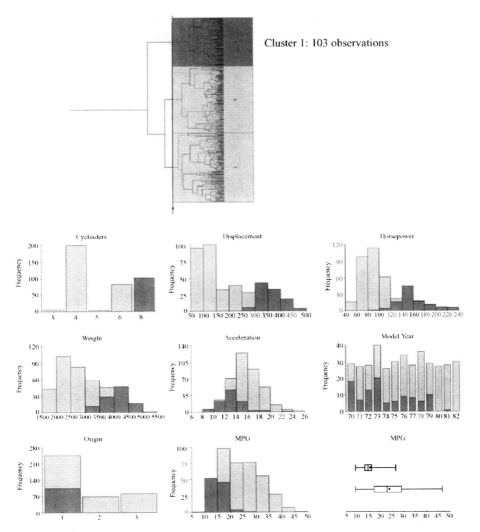

Figure 6.15.   Summary of content of Cluster 1

- **Cluster 4:** This set of 56 observations is a subset of Cluster 1 and shown in Figure 6.19. This is a set of cars with high **displacement, horsepower**, and **weight** as well as some of the lowest **acceleration** values. They were all made prior to 1976 in the US (**origin** 1). The fuel efficiency (**MPG**) of these cars is among the lowest in the data set. The range of fuel efficiency for these vehicles is 9–20 with an average of 15.34.

- **Cluster 5:** This set of 40 observations is a subset of Cluster 3 and shown in Figure 6.20. This is a set of cars with three and four **cylinders** with low **displacement, horsepower**, and **weight** and with **acceleration** values

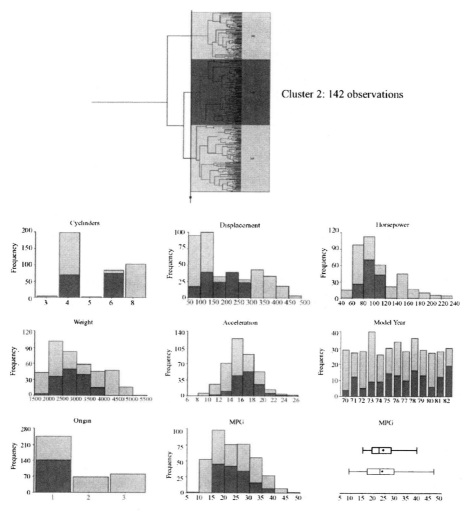

Figure 6.16. Summary of content of Cluster 2

similar to the average of the set. They were made in the late 1970s and early 1980s in Japan (**origin** 3). The range of fuel efficiency for these vehicles is 25.8–40.8 with an average of 32.46. These cars have good fuel efficiency, compared to the others in the data set.

6.2.3 K-means Clustering

Overview

K-means clustering is an example of a nonhierarchical method for grouping a data set. It groups data using a "top-down" approach since it starts with a predefined

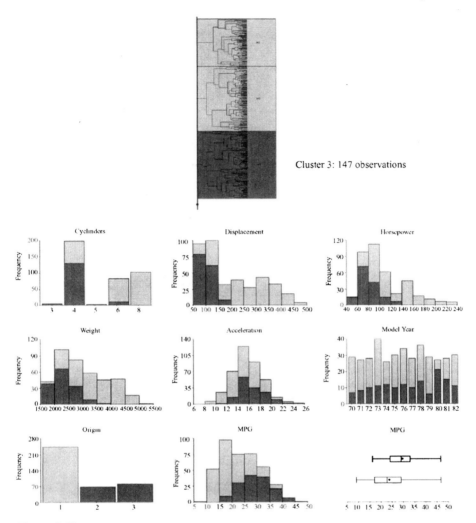

Cluster 3: 147 observations

Figure 6.17. Summary of content of Cluster 3

number of clusters and assigns observations to them. There are no overlaps in the groups, that is, all observations are assigned to a single group. This approach is computationally faster and can handle greater numbers of observations than agglomerative hierarchical clustering. However, there are a number of disadvantages:

- **Predefined number of clusters:** You must define the number of groups before creating the clusters.
- **Distorted by outliers:** When a data set contains many outliers, k-means clustering may not create an optimal grouping. This is because the outliers

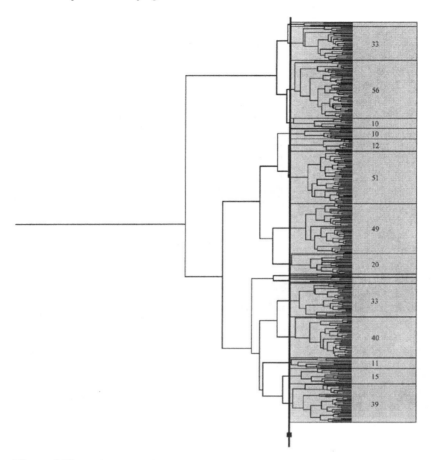

Figure 6.18. Adjusting the distance to generate 16 clusters

will be assigned to many of the allocated groups. The remaining data will then be divided across a smaller number of groups, compromising the quality of the clustering for these remaining observations.

- **No hierarchical organization:** No hierarchical organization is generated using k-means clustering.

Grouping Process

The process of generating clusters starts by defining the number of groups to create (k). The method then allocates an observation to each of these groups, usually randomly. Next, all other observations are compared to each of these allocated observations and placed in the group they are most similar to. The center point for each of these groups is then calculated. The grouping process continues by

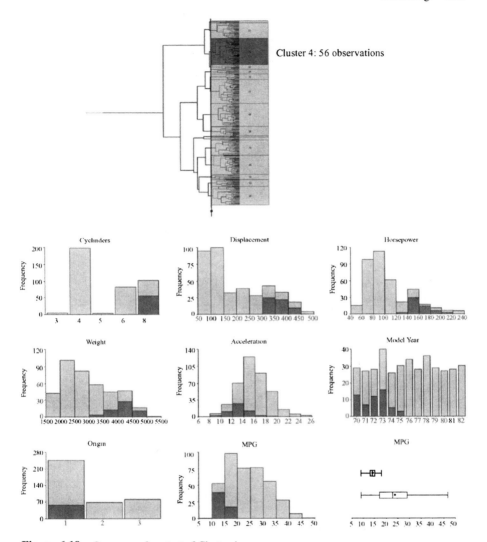

Figure 6.19. Summary of content of Cluster 4

determining the distance from all observations to these new group centers. If an observation is closer to the center of another group, it is moved to the group it is closest to. The centers of its old and new groups are now recalculated. The process of comparing and moving observations where appropriate is repeated until there is no further need to move any observations.

   To illustrate the process of clustering using k-means, a set of 11 hypothetical observations are used: a, b, c, d, e, f, g, h, i, j. These observations are shown as colored circles in Figure 6.21. It had been determined from the start that three groups should be generated. Initially, an observation is randomly assigned to each of the

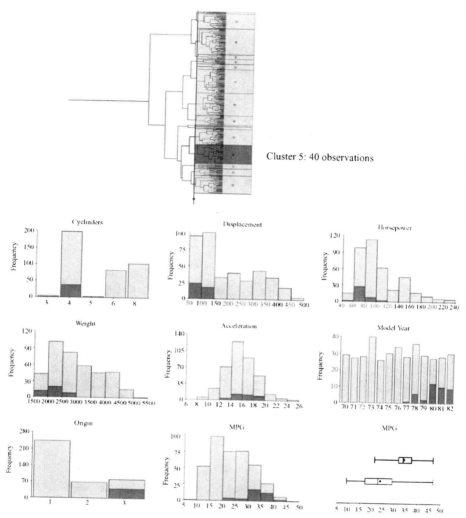

Cluster 5: 40 observations

Figure 6.20.   Summary of content of Cluster 5

three clusters as shown in step 1: c to Cluster 1, f to Cluster 2 and k to Cluster 3. Next, all remaining observations are assigned to the cluster, which they are closest to. For example, observation a is assigned to Cluster 1 since it is closer to c than f or k. Once all observations have been assigned to an initial cluster, the center of each cluster (calculation described below) is determined. Next, distances from each observation to the center of each cluster are calculated. It is determined in step 3 that observation f is closer to the center of cluster 1 than the other two clusters. Now f is moved to Cluster 1 and the centers for Cluster 1 and Cluster 2 are recalculated. This process continues until no more observations are moved between clusters, as shown in step n on the diagram.

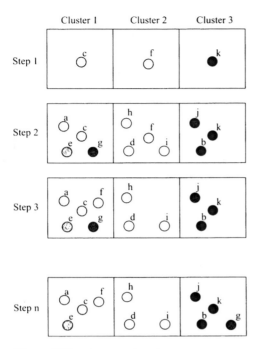

Figure 6.21. K-means clustering process

Calculating the Center of the Group

The following example will illustrate the process of calculating the center of a cluster. Table 6.9 will be grouped into three clusters using the Euclidean distance to determine similarity between observations. A single observation is randomly

Table 6.9. Table of observations to illustrate k-means clustering

Name	Variable 1	Variable 2	Variable 3	Variable 4	Variable 5
A	7	8	4	5	2
B	6	8	5	4	2
C	8	9	7	8	9
D	6	7	7	7	8
E	1	2	5	3	4
F	3	4	5	3	5
G	7	8	8	6	6
H	8	9	6	5	5
I	2	3	5	6	5
J	1	2	4	4	2
K	3	2	6	5	7
L	2	5	6	8	9
M	3	5	4	6	3
N	3	5	5	6	3

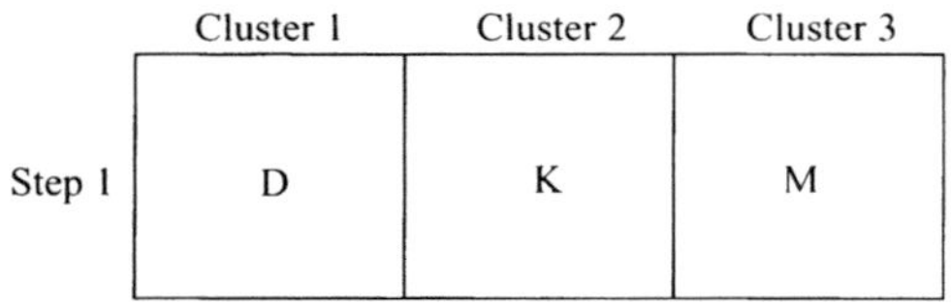

	Cluster 1	Cluster 2	Cluster 3
Step 1	D	K	M

Figure 6.22. Initial random assignment of three clusters

assigned to the three clusters as shown in Figure 6.22. All other observations are compared to the three clusters by calculating the Euclidean distance between the observations and D, K, and M. Table 6.10 shows the Euclidean distance to D, K, and M from every other observation, along with the cluster it is initially assigned to. The observations are now assigned to one of the three clusters (Figure 6.23).

Next, the center of each cluster is now calculated by taking the average value for each variable in the group as shown in Table 6.11. For example, the center of Cluster 1 is now:

{**Variable 1** = 6.2; **Variable 2** = 7.6; **Variable 3** = 6.8; **Variable 4** = 6.8;

Variable 5 = 7.4}

Each observation is now compared to the centers of each cluster. For example, A is compared to the center of Cluster 1, Cluster 2, and Cluster 3 using the Euclidean distance. We have the following Euclidean distance:

From A to the center of cluster 1: 6.4

From A to the center of cluster 2: 7.9

From A to the center of cluster 3: 3.9

Since A is still closest to Cluster 3 it remains in Cluster 3. If an observation is moved, then the centers for the two clusters affected are recalculated. The process of

Table 6.10. Observation distance to each cluster and cluster assignment

Name	Cluster 1 distance	Cluster 2 distance	Cluster 3 distance	Cluster assigned
A	7.1	9	5.2	3
B	7.1	8.5	4.9	3
C	3.2	9.4	9.5	1
E	9.3	4.2	4.9	2
F	6.9	3.6	3.9	2
G	2.8	7.6	7.1	1
H	4.7	8.8	7.1	1
I	6.8	2.8	3.2	2
J	10.2	5.8	4.2	3
L	4.8	4.8	6.7	1
N	6.6	5.2	1	3

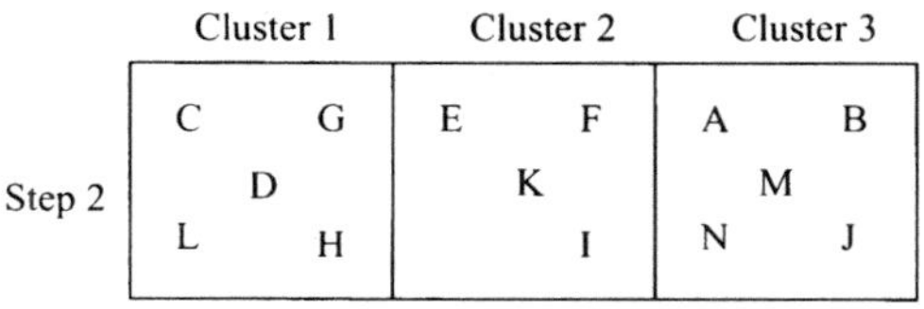

Figure 6.23.   Initial assignment of all observations

examining the observations and moving them as appropriate is repeated until no further moves are needed.

Example

A data set of 392 cars is grouped using k-means clustering. This is the same data set used in the agglomerative hierarchical clustering example. The Euclidean distance

Table 6.11.   Calculation of cluster center

	Cluster 1				
Name	**Variable 1**	**Variable 2**	**Variable 3**	**Variable 4**	**Variable 5**
C	8	9	7	8	9
D	6	7	7	7	8
G	7	8	8	6	6
H	8	9	6	5	5
L	2	5	6	8	9
Center (average)	**6.2**	**7.6**	**6.8**	**6.8**	**7.4**
	Cluster 2				
Name	**Variable 1**	**Variable 2**	**Variable 3**	**Variable 4**	**Variable 5**
E	1	2	5	3	4
F	3	4	5	3	5
I	2	3	5	6	5
K	3	2	6	5	7
Center (average)	**2.25**	**2.75**	**5.25**	**4.25**	**5.25**
	Cluster 3				
Name	**Variable 1**	**Variable 2**	**Variable 3**	**Variable 4**	**Variable 5**
A	7	8	4	5	2
B	6	8	5	4	2
J	1	2	4	4	2
M	3	5	4	6	3
N	3	5	5	6	3
Center (average)	**4**	**5.6**	**4.4**	**5**	**2.4**

was used and the number of clusters was set to 16. The same set of descriptors was used as the agglomerative hierarchical clustering example. The results are not identical; however, they produce similar clusters of observations. For example, the cluster shown in Figure 6.24 containing 35 observations is a set of similar observations to cluster 4 in the agglomerative hierarchical clustering example.

The cluster of 46 observations shown in Figure 6.25 represents a cluster with similar characteristics to Cluster 5 in the agglomerative hierarchical clustering example.

Cluster 1 (11 observations)
Cluster 2 (28 observations)
Cluster 3 (14 observations)
Cluster 4 (43 observations)
Cluster 5 (30 observations)
Cluster 6 (6 observations)
Cluster 7 (27 observations)
Cluster 8 (35 observations)
Cluster 9 (27 observations)
Cluster 10 (26 observations)
Cluster 11 (6 observations)
Cluster 12 (35 observations)
Cluster 13 (46 observations)
Cluster 14 (18 observations)
Cluster 15 (31 observations)
Cluster 16 (9 observations)

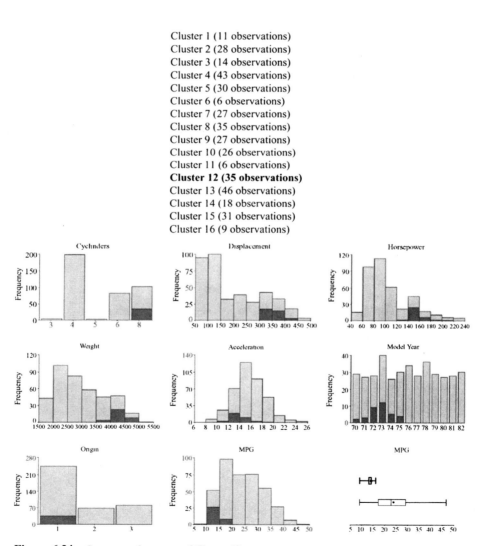

Figure 6.24.   Summary of contents of Cluster 12

Cluster 1 (11 observations)
Cluster 2 (28 observations)
Cluster 3 (14 observations)
Cluster 4 (43 observations)
Cluster 5 (30 observations)
Cluster 6 (6 observations)
Cluster 7 (27 observations)
Cluster 8 (35 observations)
Cluster 9 (27 observations)
Cluster 10 (26 observations)
Cluster 11 (6 observations)
Cluster 12 (35 observations)
Cluster 13 (46 observations)
Cluster 14 (18 observations)
Cluster 15 (31 observations)
Cluster 16 (9 observations)

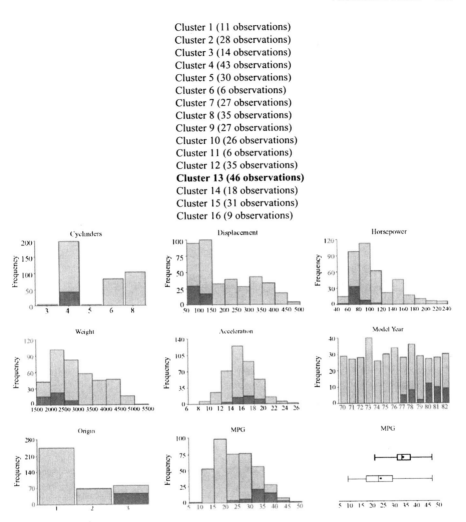

Figure 6.25. Summary of contents of Cluster 13

6.3   ASSOCIATIVE RULES

6.3.1   Overview

The associative rules method is an example of an unsupervised grouping method, that is, the goal is not used to direct how the grouping is generated. This method groups observations and attempts to understand links or associations between different attributes of the group. Associative rules have been applied in many situations, such as data mining retail transactions. This method generates rules from the groups, as the following example:

IF the customer's age is 18 AND

the customer buys paper AND

the customer buys a hole punch

THEN the customer buys a binder

The rule states that 18-year-old customers who purchase paper and a hole punch will often buy a binder at the same time. This rule would have been generated directly from a data set. Using this information the retailer may decide, for example, to create a package of products for college students.

Associative rules have a number of advantages:

- **Easy to interpret:** The results are presented in the form of a rule that is easily understood.

- **Actionable:** It is possible to perform some sort of action based on the rule. For example, the rule in the previous example allowed the retailer to market this combination of items differently.

- **Large data sets:** It is possible to use this technique with large numbers of observations.

There are three primary limitations to this method:

- **Only categorical variables:** The method forces you to either restrict your analysis to variables that are categorical or to convert continuous variable to categorical variables.

- **Time-consuming:** Generating the rules can be time-consuming for the computer; especially where a data set has many variables and/or many possible values per variable. There are ways to make the analysis run faster but they often compromise the final results.

- **Rule prioritization:** The method can generate many rules that must be prioritized and interpreted.

In this method, creating useful rules from the data is done by grouping the data, extracting rules from the groups, and then prioritizing the rules. The following sections describe the process of generating associative rules.

6.3.2   Grouping by Value Combinations

Let us first consider a simple situation concerning a shop that only sells cameras and televisions. A data set of 31,612 sales transactions is used, which contains three variables: **Customer ID**, **Gender** and **Purchase**. The variable **Gender** identifies whether the buyer is male or female. The variable **Purchase** refers to the item purchased and can only have two values, camera and television. Table 6.12 shows three rows from this table. By grouping this set of 31,612 observations, based on specific values for the variables **Gender** and **Purchase**, the groups in Table 6.13 are generated. There are eight ways of grouping this trivial

Table 6.12. Table of three example observations with three variables

Customer ID	Gender	Purchase
932085	Male	Television
596720	Female	Camera
267375	Female	Television

example based on the values for the different categories. For example, there are 7,889 observations where **Gender** is male and **Purchase** is camera.

If an additional variable is added to this data set, the number of possible groups will increase. For example, if another variable **Income** which has two values, above $50K and below $50K, is added to the table (Table 6.14), the number of groups would increase to 26 as shown in Table 6.15.

Increasing the number of variables and/or the number of possible values for each variable increases the number of groups. The number of groups may become so large that it would be impossible to generate all combinations. However, most data sets contain many possible combinations of values with zero or only a handful of observations. Techniques for generating the groups can take advantage of this fact. By increasing the minimum size of a group, fewer groups are generated and the analysis is completed faster. However, care should be taken in setting this cutoff value since no rules will be generated from any groups where the number of observations is below this cutoff. For example, if this number is set to ten, then no rules will be generated from groups containing less than ten examples. Subject matter knowledge and information generated from the data characterization phase will help in setting this value. It is a trade-off between how fast you wish the rule generation to take versus how subtle the rules need to be (i.e. rules based on a few observations).

6.3.3   Extracting Rules from Groups

Overview

So far a data set has been grouped according to specific values for each of the variables. In Figure 6.26, 26 observations (A to Z) are characterized by three

Table 6.13. Grouping by different value combinations

Group Number	Count	Gender	Purchase
Group 1	16,099	Male	-
Group 2	15,513	Female	-
Group 3	16,106	-	Camera
Group 4	15,506	-	Television
Group 5	7,889	Male	Camera
Group 6	8,210	Male	Television
Group 7	8,217	Female	Camera
Group 8	7,296	Female	Television

Table 6.14.   Table of three observations with four variables

Customer ID	Gender	Purchase	Income
932085	Male	Television	Below $50K
596720	Female	Camera	Above $50K
267375	Female	Television	Below $50K

variables, **Shape**, **Color**, and **Border**. Observation A has **Shape** = square, **Color** = white, and **Border** = thick and observation W has **Shape** = circle, **Color** = gray, and **Border** = thin.

As described in the previous section, the observations are grouped. An example grouping is shown below where:

Shape = circle,
Color = gray
Border = thick

Table 6.15.   Table showing groups by different value combinations

Group Number	Count	Gender	Purchase	Income
Group 1	16,099	Male	-	-
Group 2	15,513	Female	-	-
Group 3	16,106	-	Camera	-
Group 4	15,506	-	Television	-
Group 5	15,854	-	-	Below $50K
Group 6	15,758	-	-	Above $50K
Group 7	7,889	Male	Camera	-
Group 8	8,210	Male	Television	-
Group 9	8,549	Male	-	Below $50K
Group 10	7,550	Male	-	Above $50K
Group 11	8,217	Female	Camera	-
Group 12	7,296	Female	Television	-
Group 13	7,305	Female	-	Below $50K
Group 14	8,208	Female	-	Above $50K
Group 15	8,534	-	Camera	Below $50K
Group 16	7,572	-	Camera	Above $50K
Group 17	7,320	-	Television	Below $50K
Group 18	8,186	-	Television	Above $50K
Group 19	4,371	Male	Camera	Below $50K
Group 20	3,518	Male	Camera	Above $50K
Group 21	4,178	Male	Television	Below $50K
Group 22	4,032	Male	Television	Above $50K
Group 23	4,163	Female	Camera	Below $50K
Group 24	4,054	Female	Camera	Above $50K
Group 25	3,142	Female	Television	Below $50K
Group 26	4,154	Female	Television	Above $50K

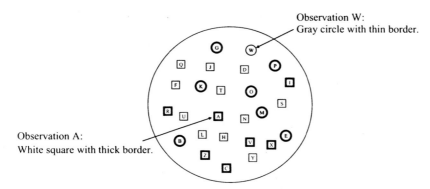

Figure 6.26. Twenty-six observations characterized by **shape, color,** and **border**

This group is shown in Figure 6.27.

The next step is to extract a rule from the group. There are three possible rules that could be pulled out from this group:

Rule 1:

IF **Color** = gray AND

Shape = circle

THEN **Border** = thick

Rule 2:

IF **Border** = thick AND

Color = gray

THEN **Shape** = circle

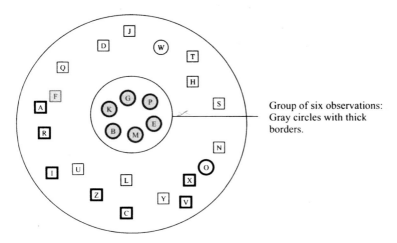

Figure 6.27. Group of six observations (gray circles with thick borders)

Rule 3:

IF **Border** = thick AND

Shape = circle

THEN **Color** = gray

We now compare each rule to the whole data set in order to prioritize the rules and three values are calculated: *support, confidence* and *lift*.

Support

The support value is another way of describing the number of observations that the rule (created from the group) maps onto, that is, the size of the group. The support is often defined as a proportion or percentage. In this example, the data set has 26 observations and the group of gray circles with a thick border is six, then the group has a support value of six out of 26 or 0.23 (23%).

Confidence

Each rule is divided into two parts. The IF-part or *antecedent* refers to the list of statements linked with AND in the first part of the rule. For example,

IF **Color** = gray AND

Shape = circle

THEN **Border** = thick

The IF-part is the list of statements **Color** = gray AND **Shape** = circle. The THEN-part of the rule or *consequence* refers to any statements after the THEN (**Border** = thick in this example).

The confidence score is a measure for how predictable a rule is. The confidence (or predictability) value is calculated using the support for the entire group divided by the support for all observations satisfied by the IF-part of the rule:

$$\text{Confidence} = \text{group support}/\text{IF-part support}$$

For example, the confidence value for Rule 1

Rule 1:

IF **Color** = gray AND

Shape = circle

THEN **Border** = thick

is calculated using the support value for the group and the support value for the IF-part of the rule (see Figure 6.28).

The support value for the group (gray circles with a thick border) is 0.23 and the support value for the IF-part of the rule (gray circles) is seven out of 26 or 0.27. To calculate the confidence, divide the support for the group by the support for the IF-part:

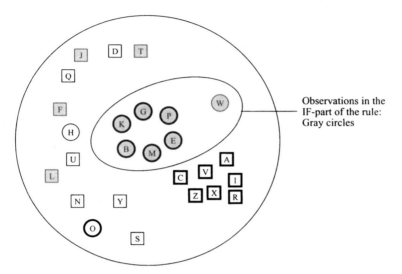

Figure 6.28. Seven observations for gray circles

$$\text{Confidence} = 0.23/0.27 = 0.85$$

Confidence values range from no confidence (0) to high confidence (1). Since a value of 0.85 is close to 1, we have a high degree of confidence in this rule. Most likely, gray circles will have thick border.

Lift

The confidence value does not indicate the strength of the association between gray circles (IF-part) and thick borders (THEN-part). The lift score takes this into account. The lift is often described as the importance of the rule. It describes the association between the IF-part of the rule and the THEN-part of the rule. It is calculated by dividing the confidence value by the support value across all observation of the THEN-part:

$$\text{Lift} = \text{confidence}/\text{THEN-part support}$$

For example, the lift for Rule 1

Rule 1:

IF **Color** = gray AND

Shape = circle

THEN **Border** = thick

is calculated using the confidence and the support for the THEN-part of the rule (see Figure 6.29). The confidence for rule 1 is calculated as 0.85 and the support for the THEN-part of the rule (thick borders) is 13 out of 26 or 0.5. To calculate the

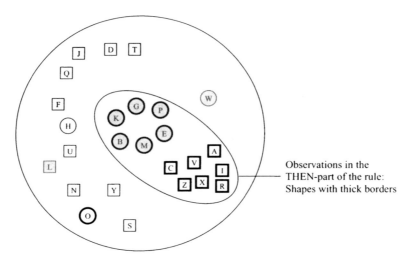

Figure 6.29.   Thirteen observations for thick border objects

lift value, the confidence is divided by the support value for the THEN-part of the rule:

$$\text{Lift} = 0.85/0.5 = 1.7$$

Lift values greater than 1 indicate a positive association and lift values less than 1 indicate a negative association.

Figure 6.30 is used to determine the confidence and support for all three potential rules.

The following shows the calculations for support, confidence and lift for the three rules:

Rule 1:
Support $= 6/26 = 0.23$
Confidence $= 0.23/(7/26) = 0.85$
Lift $= 0.85/(13/26) = 1.7$

Rule 2:
Support $= 6/26 = 0.23$
Confidence $= 0.23/(6/26) = 1$
Lift $= 1/(9/26) = 2.9$

Rule 3:
Support $= 6/26 = 0.23$
Confidence $= 0.23/(7/26) = 0.85$
Lift $= 0.85 / (10 / 26) = 2.2$

Rule 1:
IF **Color** = gray AND
Shape = circle
THEN **Border** = thick

Rule 2:
IF **Border** = thick AND
Color = gray
THEN **Shape** = circle

Rule 3:
IF **Border** = thick AND
Shape = circle
THEN **Color** = gray

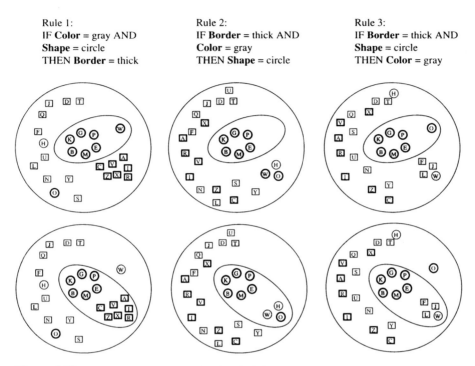

Figure 6.30. Separating objects for each rule calculation

The values are summarized in Table 6.16.

Rule 2 would be considered the most interesting because of the confidence score of 1 and the high positive lift score indicating that shapes that are gray with a thick border are likely to be circles.

6.3.4 Example

In this example, we will compare two rules generated from the Adult data available from Newman (1998). This is a set of income data with the following variables along with all possible values shown in parenthesis:

- **Class of work** (Private, Self-emp-not-inc, Self-emp-inc, Federal-gov, Local-gov, State-gov, Without-pay, Never-worked)

Table 6.16. Summary of support, confidence, and lift for the three rules

	Rule 1	Rule 2	Rule 3
Support	0.23	0.23	0.23
Confidence	0.85	1.0	0.85
Lift	1.7	2.9	2.2

- **Education** (Bachelors, Some-college, 11th, HS-grad, Prof-school, Assoc-acdm, Assoc-voc, 9th, 7th–8th, 12th, Masters, 1st–4th, 10th, Doctorate, 5th–6th, Preschool)
- **Income** ($> 50K$, $\leq 50K$)

There are 32,561 observations and using the associative rule method, many rules were identified. For example,

Rule 1

IF Class of work is Private and
Education is Doctorate
THEN Income is $\leq 50K$

Rule 2

IF Class of work is Private and
Education is Doctorate
THEN Income is $> 50K$

Here is a summary of the counts:

Class of work is private: 22,696 observations.
Education is Doctorate: 413 observations.
Class of work is private and **Education** is Doctorate: 181 observations.
Income is less than or equal to 50K: 24,720 observations.
Income is greater than 50K: 7,841 observations.

Table 6.17 shows the information calculated for the rules. Of the 181 observations where **Class of work** is private and **Education** is Doctorate, 132 (73%) of those observations also had **Income** greater than 50K. This is reflected in the much higher confidence score for rule 2 (0.73) compared to rule 1 (0.27). Over the entire data set of 32,561 observations there are about three times the number of observations where income is less than or equal to 50K as compared to observations where the income is greater than 50K. The lift term takes into consideration the relative frequency of the THEN-part of the rule. Hence, the lift value for rule 2 is considerably higher (3.03) than the lift value for rule 1. Rule 2 has both a good confidence and lift value, making it an interesting rule. Rule 1 has both a poor confidence and lift value. The following examples illustrate some other rules generated:

Table 6.17. Summary of scores for two rules

	Rule 1	Rule 2
Count	49	132
Support	0.0015	0.0041
Confidence	0.27	0.73
Lift	0.36	3.03

Rule 3

IF **Class of work** is State-gov and
Education is 9th
THEN **Income** is $\leq$ 50K
(Count: 6; Support: 0.00018; Confidence: 1; Lift: 1.32)

Rule 4

IF **Class of work** is Self-emp-inc and
Education is Prof-school
THEN **Income** is >50 K
(Count: 78; Support: 0.0024 Confidence: 0.96; Lift: 4)

Rule 5

IF **Class of work** is Local-gov and
Education is 12th
THEN **Income** is $\leq$50 K
(Count: 17; Support: 0.00052; Confidence: 0.89; Lift: 1.18)

6.4 DECISION TREES

6.4.1 Overview

It is often necessary to ask a series of questions before coming to a decision. The answers to one question may lead to another question or may lead to a decision being reached. For example, you may visit a doctor and your doctor may ask you to describe your symptoms. You respond by saying you have a stuffy nose. In trying to diagnose your condition the doctor may ask you further questions such as whether you are suffering from extreme exhaustion. Answering yes would suggest you have the flu, whereas answering no would suggest you have a cold. This line of questioning is common to many decision making processes and can be shown visually as a decision tree, as shown in Figure 6.31.

Decision trees are often generated by hand to precisely and consistently define a decision making process. However, they can also be generated automatically from the data. They consist of a series of decision points based on certain variables. Figure 6.32

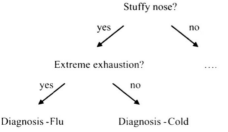

Figure 6.31. Decision tree for the diagnosis of colds and flu

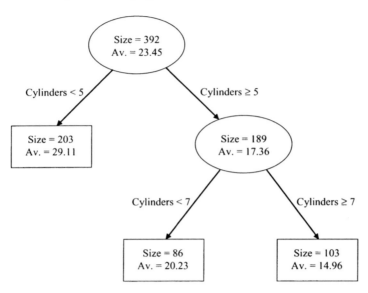

Figure 6.32. Decision tree generated from a data set of cars

illustrates a simple decision tree. This decision tree was generated based on a data set of cars which included a variable number of cylinders (**Cylinders**) along with the car fuel efficiency (**MPG**). The decision tree attempts to group cars based on the number of cylinders (**Cylinders**) in order to classify the observations according to their fuel efficiency. At the top of the tree is a node representing the entire data set of 392 observations (Size = 392). The data set is initially divided into two subsets, on the left of the Figure is a set of 203 cars (i.e. Size = 203) where the number of cylinders is less than five. How this division was determined will be described later in this section. Cars with less than five cylinders are grouped together as they generally have good fuel efficiency with an average **MPG** value of 29.11. The remaining 189 cars are further grouped into a set of 86 cars where the number of cylinders is less than seven. This set does not include any cars with less than five cylinders since they were separated earlier. These cars are grouped as they generally have reasonable fuel efficiency with an average **MPG** value of 20.23. The remaining group is a set of cars where the number of cylinders is greater than seven and these have poor fuel efficiency with an average **MPG** value of 14.96.

In contrast with clustering or association rules, decision trees are an example of a supervised method. In this example, the data set was classified into groups using the variable **MPG** to guide how the tree was constructed. Figure 6.33 illustrates how the tree was put together, guided by the data. A histogram of the **MPG** response data is shown alongside the nodes used to classify the vehicles. The overall shape of the histogram displays the frequency distribution for the **MPG** variable. The highlighted frequency distribution is the subset within the node. The frequency distribution for the node containing 203 observations shows a set biased toward good fuel efficiency, whereas the 103 observations highlighted illustrate a set biased towards poor fuel

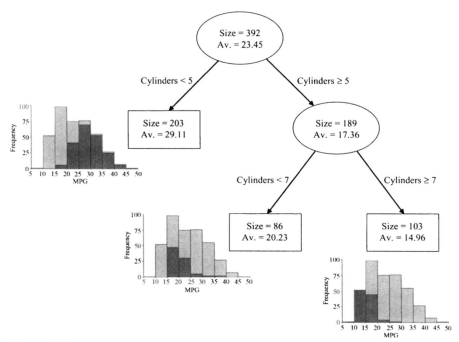

Figure 6.33. Decision tree illustrating the use of a response variable (**MPG**) to guide tree generation

efficiency. The **MPG** variable has not been used in any of the decision points, only the number of cylinders. This is a trivial example, but it shows how a data set can be divided into regions using decision trees.

There are many reasons to use decision trees:

- **Easy to understand:** Decision trees are widely used to explain how decisions are reached based on multiple criteria.

- **Categorical and continuous variables:** Decision trees can be generated using either categorical data or continuous data.

- **Complex relationships:** A decision tree can partition a data set into distinct regions based on ranges or specific values.

The disadvantages of decision trees are:

- **Computationally expensive:** Building decision trees can be computation-ally expensive, particularly when analyzing a large data set with many continuous variables.

- **Difficult to optimize:** Generating a useful decision tree automatically can be challenging, since large and complex trees can be easily generated. Trees that are too small may not capture enough information. Generating the 'best' tree through optimization is difficult. At the end of this chapter, a series of references to methods of decision tree optimization can be found.

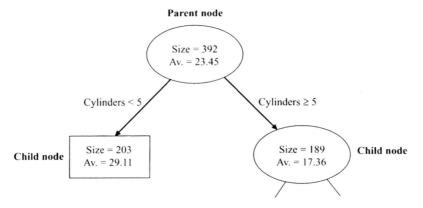

Figure 6.34.   Relationship between parent and child node

6.4.2   Tree Generation

A tree is made up of a series of decision points, where the entire set of observations or a subset of the observations is split based on some criteria. Each point in the tree represents a set of observations and is called a *node*. The relationship between two nodes that are joined is defined as a parent-child relationship. The larger set which will be divided into two or more smaller sets is called the *parent node*. The nodes resulting from the division of the parent are called *child nodes* as shown in Figure 6.34. A child node with no more children (no further division) is called a *leaf node* and shown in Figure 6.35.

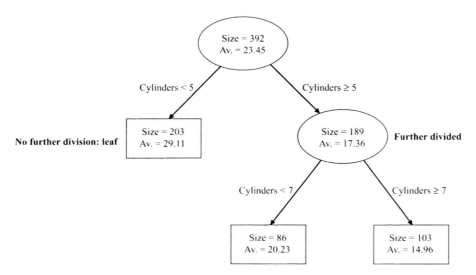

Figure 6.35.   Leaf and nonleaf nodes

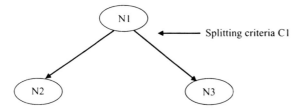

Figure 6.36. Splitting a set of observations into two groups

A table of data is used to generate a decision tree where certain variables are assigned as descriptors and one variable is assigned to be the response. The descriptors will be used to build the tree, that is, these variables will divide the data set. The response will be used to guide which descriptors are selected and at what value the split is made. A decision tree splits the data set into smaller and smaller sets. The head (or top) of the tree is a node containing all observations. Based on some criteria, the observations are split resulting in usually two new nodes, each node representing a smaller set of observations, as shown in Figure 6.36. Node N1 represents all observations. By analyzing all descriptor variables and examining many splitting points for each variable, an initial split is made based on some criteria (C1). The data set represented at node N1 is now divided into a set N2 that meets criteria C1 and a set N3 that does not satisfy the criteria.

The process of examining the variables to determine a criterion for splitting is repeated for all subsequent nodes. However, a condition should be specified for ending this repetitive process. For example, the process can stop when the size of the subset is less than a predetermined value. In Figure 6.37, each of the two newly created subset (N2 and N3) are examined in turn to determine if they should be further split or if the splitting should stop.

In Figure 6.38, the subset at node N2 is examined to determine if the splitting should stop. Here, the condition for stopping splitting is not met and hence the subset is to be split further. Again, all the variables assigned as descriptors are considered along with many alternatives values to split on. The best criterion is selected and the data set is again divided into two sets, represented by N4 and N5. Set N4 represents a set of observations that satisfy the splitting criteria (C2) and node N5 is the remaining set of observations. Next, node N3 is examined and in this case, the condition to stop splitting is met and the subset represented by node N3 is not divided further.

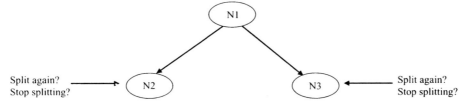

Figure 6.37. Evaluating whether to continue to grow the tree

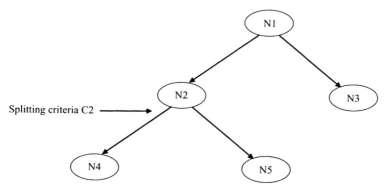

Figure 6.38.   Tree further divided

6.4.3   Splitting Criteria

Dividing Observations

It is common for the split at each level to be a two-way split. There are methods that split more than two ways. However, care should be taken using these methods since splitting the set in many ways early in the construction of the tree may result in missing interesting relationships that become exposed as the tree growing process continues. Figure 6.39 illustrates the two alternatives.

Any variable type can be split using a two-way split:

- **Dichotomous:** Variables with two values are the most straightforward to split since each branch represents a specific value. For example, a variable **Temperature** may have only two values, hot and cold. Observations will be split based on those with hot and those with cold temperature values.

- **Nominal:** Since nominal values are discrete values with no order, a two-way split is accomplished with one subset being comprised of a set of observations that equal a certain value and the other subset being those observations that do not equal that value. For example, a variable **Color** that can take the values red, green, blue, and black may be split two-ways. Observations, for

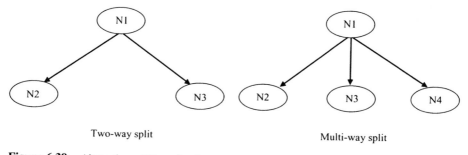

Two-way split   Multi-way split

Figure 6.39.   Alternative splitting of nodes

example, which have **Color** equaling red generate one subset and those not equaling red creating the other subset, that is, green, blue and black.

- **Ordinal:** In the case where a variable's discrete values are ordered, the resulting subsets may be made up of more than one value, as long as the ordering is retained. For example, a variable **Quality** with possible values low, medium, high, and excellent may be split in four possible ways. For example, observations equaling low or medium in one subset and observations equaling high and excellent in another subset. Another example is where low values are in one set and medium, high, and excellent values are in the other set.

- **Continuous:** For variables with continuous values to be split two-ways, a specific cutoff value needs to be determined, where on one side of the split are values less than the cutoff and on the other side of the split are values greater than or equal to the cutoff. For example, a variable **Weight** which can take any value between 0 and 1,000 with a selected cutoff of 200. The first subset would be those observations where the **Weight** is below 200 and the other subset would be those observations where the **Weight** is greater than or equal to 200.

Figure 6.40 illustrates how the different variable types can be used as splitting criteria in a two-way split.

A splitting criterion has two components: (1) the variable to split on and (2) values of the variable to split on. To determine the best split, all possible splits of all variables must be considered. Since it is necessary to rank the splits, a score should be calculated for each split. There are many ways to rank the split. The following describes two approaches for prioritizing splits, based on whether the response is categorical or continuous.

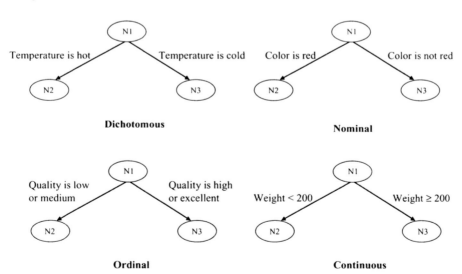

Figure 6.40. Splitting examples based on variable type

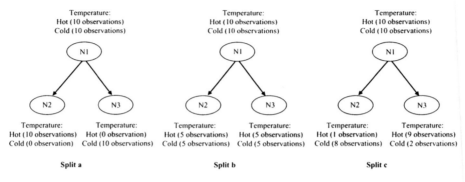

Figure 6.41.   Evaluating splits based on categorical response data

Scoring Splits for Categorical Response Variables

To illustrate how to score splits when the response is a categorical variable, three splits (split a, split b, split c) for a set of observations are shown in Figure 6.41. The objective for an optimal split is to create subsets which results in observations with a single response value. In this example, there are 20 observations prior to splitting. The response variable (**Temperature**) has two possible values, hot and cold. Prior to the split, the response has an even distribution with the number of observations where the **Temperature** equals hot is ten and with the number of observations where the **Temperature** equals cold is also ten.

Different criteria are considered for splitting these observations which results in different distributions of the response variables for each subset (N2 and N3):

- **Split a:** Each subset contains ten observations. All ten observations in N2 have hot temperature values, whereas the ten observations in node N3 are all cold.

- **Split b:** Again each subset (N2 and N3) contains ten observations. However, in this example there is an even distribution of hot and cold values in each subset.

- **Split c:** In this case the splitting criterion results in two subsets where node N2 has nine observations (one hot and eight cold) and node N3 has 11 observations (nine hot and two cold).

Split a is the best split since each node contains observations where the response is one or the other category. **Split b** results in the same even split of hot and cold values (50% hot, 50% cold) in each of the resulting nodes (N2 and N3) and would not be considered a good split. **Split c** is a good split; however, this split is not so clean as **split a** since there are values of both hot and cold in both subsets. The proportion of hot and cold values is biased, in node N2 towards cold values and in N3 towards hot values. When determining the best splitting criteria, it is therefore important to determine how clean each split is, based on the proportion of the different categories of the response variable (or *impurity*). As the tree is being generated, it is desirable to decrease the level of impurity until, in an ideal situation, there is only one response value at a terminal node.

Table 6.18. Entropy scores according to different splitting
criteria

Scenario	Response values hot	cold	Entropy
Scenario 1	0	10	0
Scenario 2	1	9	0.469
Scenario 3	2	8	0.722
Scenario 4	3	7	0.881
Scenario 5	4	6	0.971
Scenario 6	5	5	1
Scenario 7	6	4	0.971
Scenario 8	7	3	0.881
Scenario 9	8	2	0.722
Scenario 10	9	1	0.469
Scenario 11	10	0	0

There are three primary methods for calculating impurity: *misclassification*, *Gini*, and *entropy*. In the following examples the entropy calculation will be used; however, the other methods give similar results. To illustrate the use of the entropy calculation, a set of ten observations with two possible response values (hot and cold) are used (Table 6.18). All possible scenarios for splitting this set of ten observations are shown: Scenario 1 through 11. In scenario 1, all ten observations have value cold whereas in scenario 2, one observation has value hot and nine observations have value cold. For each scenario, an entropy score is calculated. Cleaner splits result in lower scores. In scenario 1 and scenario 11, the split cleanly breaks the set into observations with only one value. The score for these scenarios is 0. In scenario 6, the observations are split evenly across the two values and this is reflected in a score of 1. In other cases, the score reflects how well the two values are split.

The formula for entropy is:

$$Entropy(S) = -\sum_{i=1}^{c} p_i \log_2 p_i$$

The entropy calculation is performed on a set of observations S. p_i refers to the fraction of the observations that belong to a particular values. For example, for a set of 100 observations where the response variable is **Temperature** and 60 observations had hot values while 40 observations had cold values, then the p_{hot} would be 0.6 and the p_{cold} would be 0.4. The value c is the number of different values that the response variable can take. When $p_i = 0$, then the value for $0 \log_2(0) = 0$.

The example shown in Figure 6.41 will be used to illustrate our point. Values for entropy are calculated for the three splits:

split a

Entropy (N1) $= -(10/20) \log_2 (10/20) - (10/20) \log_2 (10/20) = 1$

Entropy (N2) $= -(10/10) \log_2 (10/10) - (0/10) \log_2 (0/10) = 0$

Entropy(N3) $= -(0/10) \log_2 (0/10) - (10/10) \log_2 (10/10) = 0$

split b

Entropy (N1) $= -(10/20) \log_2 (10/20) - (10/20) \log_2 (10/20) = 1$

Entropy (N2) $= -(5/10) \log_2 (5/10) - (5/10) \log_2 (5/10) = 1$

Entropy (N3) $= -(5/10) \log_2 (5/10) - (5/10) \log_2 (5/10) = 1$

split c

Entropy (N1) $= -(10/20) \log_2 (10/20) - (10/20) \log_2 (10/20) = 1$

Entropy (N2) $= -(1/9) \log_2 (1/9) - (8/9) \log_2 (8/9) = 0.503$

Entropy (N3) $= -(9/11) \log_2 (9/11) - (2/11) \log_2 (2/11) = 0.684$

In order to determine the best split, we now need to calculate a ranking based on how cleanly each split separates the response data. This is calculated on the basis of the impurity before and after the split. The formula for this calculation, *Gain*, is shown below:

$$Gain = Entropy(parent) - \sum_{j=1}^{k} \frac{N(v_j)}{N} Entropy(v_j)$$

N is the number of observations in the parent node, k is the number of possible resulting nodes and $N(v_j)$ is the number of observations for each of the j child nodes. v_j is the set of observations for the jth node. It should be noted that the Gain formula can be used with other impurity methods by replacing the entropy calculation.

In the example described throughout this section, the gain values are calculated and shown in Figure 6.42.

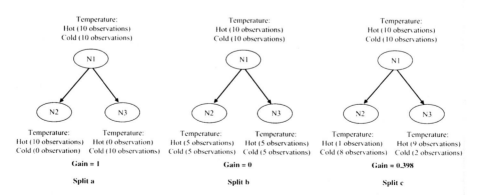

Figure 6.42.   Calculation of gain for each split

$$Gain(split\ a) = 1 - \{[(10/20)0] + [(10/20)0]\} = 1$$
$$Gain(split\ b) = 1 - \{[(10/20)1] + [(10/20)1]\} = 0$$
$$Gain(split\ c) = 1 - \{[(9/20)0.503] + [(11/20)0.684]\} = 0.397$$

The criterion used in **split a** is selected as the best splitting criteria.

During the tree generation process all possible splitting values for all descriptor variables are calculated and the best splitting criterion is selected.

Scoring Splits for Continuous Response Variables

When the response variable is continuous, one popular method for ranking the splits is to use the *sum of the squares of error* (*SSE*). The resulting split should ideally result in sets where the response values are close to the mean of the group. The lower the *SSE* value for the group, the closer the group values are to the mean of the set. For each potential split, a *SSE* value is calculated for each resulting node. A score for the split is calculated by summing the *SSE* values of each resulting node. Once all splits for all variables are computed, then the split with the lowest score is selected.

The formula for *SSE* is:

$$SSE = \sum_{i=1}^{n} (y_i - \bar{y})^2$$

For a subset of n observations, the *SSE* value is computed where y_i is the individual value for the response, and $\bar{y}$ is the average value for the subset. To illustrate, Table 6.19 is processed to identify the best split. The variable **Weight** is assigned as a descriptor and **MPG** will be used as the response variable. A series of splitting point values for the variable **Weight** will be used: 1693, 1805, 1835, 3225, 4674, 4737, and 4955. These points are the midpoints between each pair of values and were selected because they divided the data set into all possible two-ways splits, as shown in Figure 6.43. In this example, we will only calculate a score for splits which result in three or more observations, that is Split 3, Split 4, and Split 5. The **MPG** response variable is used to calculate the score.

Table 6.19.   Table of eight observations with values for two variables

	Weight	MPG
A	1,835	26
B	1,773	31
C	1,613	35
D	1,834	27
E	4,615	10
F	4,732	9
G	4,955	12
H	4,741	13

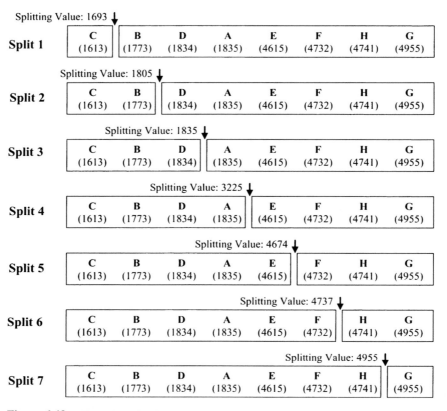

Figure 6.43.   Illustration of splitting points

Split 3:

For subset where **Weight** is less than 1835 (C, B, D):

$$\text{Average} = (35 + 31 + 27)/3 = 31$$
$$\text{SSE} = (35 - 31)^2 + (31 - 31)^2 + (27 - 31)^2 = 32$$

For subset where **Weight** is greater than or equal to 1835 (A, E, F, H, G):

$$\text{Average} = (26 + 10 + 9 + 13 + 12)/5 = 14$$
$$\text{SSE} = (26-14)^2 + (10-14)^2 + (9-14)^2 + (13-14)^2 + (12-14)^2 = 190$$

Split score $= 32 + 190 = 222$

Split 4:

For subset where **Weight** is less than 3225 (C, B, D, A):

$$\text{Average} = (35 + 31 + 27 + 26)/4 = 29.75$$

$$SSE = (35 - 29.75)^2 + (31 - 29.75)^2 + (27 - 29.75)^2 + (26 - 29.75)^2 = 50.75$$

For subset where **Weight** is greater than or equal to 3225 (E, F, H, G):

$$Average = (10 + 9 + 13 + 12)/4 = 11$$

$$SSE = (10 - 11)^2 + (9 - 11)^2 + (13 - 11)^2 + (12 - 11)^2 = 10$$

Split score $= 50.75 + 10 = 60.75$

Split 5:

For subset where **Weight** is less than 4674 (C, B, D, A, E):

$$Average = (35 + 31 + 27 + 26 + 10)/5 = 25.8$$
$$SSE = (35 - 25.8)^2 + (31 - 25.8)^2 + (27 - 25.8)^2 + (26 - 25.8)^2$$
$$+ (10 - 25.8)^2 = 362.8$$

For subset where **Weight** is greater than or equal to 4674 (F, H, G):

$$Average = (9 + 13 + 12)/3 = 11.33$$
$$SSE = (9 - 11.33)^2 + (13 - 11.33)^2 + (12 - 11.33)^2 = 8.67$$

Split score $= 362.8 + 8.67 = 371.47$

In this example, Split 4 has the lowest score and would be selected as the best split.

6.4.4 Example

In the following example, a set of 392 cars is analyzed using a decision tree. Two variables were used as descriptors: **Horsepower** and **Weight**; **MPG** (miles per gallon) was used as the response. A decision tree (Figure 6.44) was automatically generated using a 40 nodes minimum as a terminating criterion.

The leaf nodes of the tree can be interpreted using a series of rules. The decision points that are crossed in getting to the node are the rule conditions. The average **MPG** value for the leaf nodes will be interpreted here as low (less than 22), medium (22–26), and high (greater than 26). The following two example rules can be extracted from the tree:

Node A:
IF **Horsepower** < 106 AND **Weight** < 2067.5
THEN **MPG** is high

Node B:
IF **Horsepower** < 106 AND **Weight** 2067.5–2221.5
THEN **MPG** is high

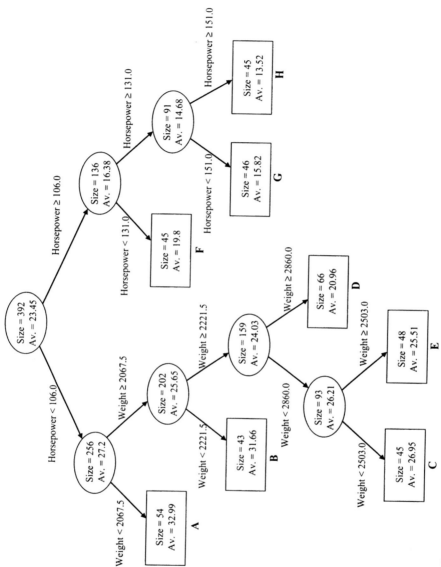

Figure 6.44. Decision tree generated using **Horsepower** and **Weight** at splitting points and guided by **MPG**

In addition to grouping data sets, decision trees can also be used in making predictions and this will be seen again in Section 7.4.

6.5 SUMMARY

Table 6.20 summarizes the different grouping methods described in this chapter.

6.6 EXERCISES

Patient data was collected concerning the diagnosis of cold or flu (Table 6.21).

1. Calculate the Euclidean distance (replacing None with 0, Mild with 1 and Severe with 2) using the variables: **Fever, Headaches, General aches, Weakness, Exhaustion, Stuffy nose, Sneezing, Sore throat, Chest discomfort,** for the following pairs of patient observations from Table 6.21:
   a. 1326 and 398
   b. 1326 and 1234
   c. 6377 and 2662
2. The patient observations described in Table 6.21 are being clustered using agglomerative hierarchical clustering. The Euclidean distance is used to calculate the distance between observations using the following variables: **Fever, Headaches, General aches, Weakness, Exhaustion, Stuffy nose, Sneezing, Sore throat, Chest discomfort** (replacing None with 0, Mild with 1 and Severe with 2). The average linkage joining rule is being used to create the hierarchical clusters. During the clustering process observations 6377 and 2662 are already grouped together. Calculate the distance from observation 398 to this group.
3. A candidate rule has been extracted using the associative rule method from Table 6.1:
   If **Exhaustion** = None AND
   Stuffy node = Severe
   THEN **Diagnosis** = cold

   Calculate the support, confidence, and lift for this rule.
4. Table 6.21 is to be used to build a decision tree to classify whether a patient has a cold or flu. As part of this process the **Fever** column is being considered as a splitting point. Two potential splitting values are being considered:
   a. Where the data is divided into two sets where (1) **Fever** is none and (2) **Fever** is mild and severe.
   b. Where the data is divided into two sets where (1) **Fever** is severe and (2) **Fever** is none and mild.

   Calculate, using the entropy impurity calculation, the gain for each of these splits.

Table 6.20. Summary of grouping methods described in this chapter

	Criteria	Data	Supervised/ unsupervised	Size	Time to compute	Overlapping groups	Comments
Agglomerative hierarchical clustering	Distances	Any	Unsupervised	Small	Slow	No	Hierarchical relationships Flexible cluster definitions
K-means clustering	Distances	Any	Unsupervised	Any	Faster than hierarchical methods	No	Predefined number of groups
Association rules	Categorical values	Categorical	Unsupervised	Large	Dependent on parameters	Yes	Presents associations between values
Decision trees	Categorical values or ranges	Any	Supervised	Large	Dependent on variables used	No overlapping in terminal nodes	Organizes the data into a tree Create rules from terminal nodes

Table 6.21.   Table of patient records

Patient id	Fever	Head-aches	General aches	Weak-ness	Exha-ustion	Stuffy nose	Sneezing	Sore throat	Chest disco-mfort	Diagn-osis
1326	None	Mild	None	None	None	Mild	Severe	Severe	Mild	Cold
398	Severe	Severe	Severe	Severe	Severe	None	None	Severe	Severe	Flu
6377	Severe	Severe	Mild	Severe	Severe	Severe	None	Severe	Severe	Flu
1234	None	None	None	Mild	None	Severe	None	Mild	Mild	Cold
2662	Severe	Severe	Mild	Severe	Severe	Severe	None	Severe	Severe	Flu
9477	None	None	None	Mild	None	Severe	Severe	Severe	None	Cold
7286	Severe	Severe	Severe	Severe	Severe	None	None	None	Severe	Flu
1732	None	None	None	None	None	Severe	Severe	None	Mild	Cold
1082	None	Mild	Mild	None	None	Severe	Severe	Severe	Severe	Cold
1429	Severe	Severe	Severe	Mild	Mild	None	Severe	None	Severe	Flu
14455	None	None	None	Mild	None	Severe	Mild	Severe	None	Cold
524	Severe	Mild	Severe	Mild	Severe	None	Severe	None	Mild	Flu
1542	None	None	Mild	Mild	None	Severe	Severe	Severe	None	Cold
8775	Severe	Severe	Severe	Severe	Mild	None	Severe	Severe	Severe	Flu
1615	Mild	None	None	Mild	None	Severe	None	Severe	Mild	Cold
1132	None	None	None	None	None	Severe	Severe	Severe	Severe	Cold
4522	Severe	Mild	Severe	Mild	Mild	None	None	None	Severe	Flu

6.7 FURTHER READING

For additional information on general data mining grouping approaches and outlier detection, see Witten (2000), Han (2005), and Hand (2001). Everitt (2001) provides further details concerning similarity methods and clustering approaches, and Quilan (1993) gives a comprehensive analysis of decision trees. In addition, Hastie (2003) covers in detail additional grouping approaches.

Chapter 7

Prediction

7.1 INTRODUCTION

7.1.1 Overview

Predictive models are used in many situations where an estimate or forecast is required, for example, to project sales or forecast the weather. A predictive model will calculate an estimate for one or more variables (responses), based on other variables (descriptors). For example, a data set of cars is used to build a predictive model to estimate car fuel efficiency (**MPG**). A portion of the observations are shown in Table 7.1. A model to predict car fuel efficiency was built using the **MPG** variable as the response and the variables **Cylinders**, **Displacement**, **Horsepower**, **Weight**, and **Acceleration** as descriptors. Once the model has been built, it can be used to make predictions for car fuel efficiency. For example, the observations in Table 7.2 could be presented to the model and the model would predict the **MPG** column.

A predictive model is some sort of mathematical equation or process that takes the descriptor variables and calculates an estimate for the response or responses. The model attempts to understand the relationship between the input descriptor variables and the output response variables; however, it is just a representation of the relationship. Rather than thinking any model generated as correct or not correct, it may be more useful to think of these models as useful or not useful to what you are trying to accomplish.

Predictive models have a number of uses including:

- **Prioritization:** Predictive models can be used to swiftly profile a data set that needs to be prioritized. For example, a credit card company may build a predictive model to estimate which individuals would be the best candidates for a direct mailing campaign. This model could be run over a database of millions of potential customers to identify a subset of the most promising customers. Alternatively, a team of scientists may be about to conduct a costly experiment and they wish to prioritize which alternative experiments have the greatest chance of success. To this end, a prediction model is built to

Table 7.1.   Table of cars with known MPG values

Names	Cylinders	Displacement	Horsepower	Weight	Acceleration	MPG
Chevrolet Chevelle Malibu	8	307	130	3,504	12	18
Buick Skylark 320	8	350	165	3,693	11.5	15
Plymouth Satellite	8	318	150	3,436	11	18
AMC Rebel SST	8	304	150	3,433	12	16
Ford Torino	8	302	140	3,449	10.5	17
Ford Galaxie 500	8	429	198	4,341	10	15
Chevrolet Impala	8	454	220	4,354	9	14
Plymouth Fury III	8	440	215	4,312	8.5	14
Pontiac Catalina	8	455	225	4,425	10	14
AMC Ambassador DPL	8	390	190	3,850	8.5	15

test the various experimental scenarios. The experiments predicted to have the highest chance of success will be tested first.

• **Decision support:** Prediction models can also be used to estimate future events so that appropriate actions can be taken. For example, prediction models are used to forecast adverse weather conditions and that information is used to trigger events such as alerting emergency services to prepare any affected neighborhoods.

• **Understanding:** Since predictive models attempt to understand the relationship between the input descriptor variables and the output response variables,

Table 7.2.   Table of cars where MPG is to be predicted

Names	Cylinders	Displacement	Horsepower	Weight	Acceleration	MPG
Dodge Challenger SE	8	383	170	3,563	10	
Plymouth Cuda 340	8	340	160	3,609	8	
Chevrolet Monte Carlo	8	400	150	3,761	9.5	
Buick Estate Wagon (SW)	8	455	225	3,086	10	
Toyota Corona Mark II	4	113	95	2,372	15	
Plymouth Duster	6	198	95	2,833	15.5	
AMC Hornet	6	199	97	2,774	15.5	
Ford Maverick	6	200	85	2,587	16	
Datsun Pl510	4	97	88	2,130	14.5	
Volkswagen 1131 Deluxe Sedan	4	97	46	1,835	20.5	

Table 7.3.   Different classification and regression methods

Classification	Regression
Classification trees	Regression trees
k-Nearest Neighbors	k-Nearest Neighbors
Logistic regression	Linear regressions
Naïve Bayes classifiers	Neural networks
Neural networks	Nonlinear regression
Rule-based classifiers	Partial least squares
Support vector machines	Support vector machines

they can be helpful beyond just calculating estimates. For example, if a prediction model was built based on a set of scientific experiments, the model will be able to suggest what variables are most important and how they contribute to the problem under investigation.

There are many methods for building prediction models and they are often characterized based on the response variable. When the response is a categorical variable, the model is called a *classification* model. When the response is a continuous variable, then the model is a *regression* model. Table 7.3 summarizes some of the methods available.

There are two distinct phases, each with a unique set of processes and issues to consider:

- **Building:** The prediction model is built using existing data called the *training set*. This training set contains examples with values for the descriptor and response variables. The training set is used to determine and quantify the relationships between the input descriptors and the output response variables. This set will be divided into observations used to build the model and assess the quality of any model built.

- **Applying:** Once a model has been built, a data set with no output response variables can be fed into this model and the model will produce an estimate for this response. A measure that reflects the confidence in this prediction is often calculated along with an explanation of how the value was generated.

7.1.2 Classification

A classification model is built to assign observations into two or more distinct categories. For example, a classification model may be built to estimate whether a customer will buy or will not buy a particular product. In another example, a classification model may be built to predict whether drilling in a particular area will result in finding oil or not.

In Figure 7.1, a set of observations are plotted using two variables. The points in light gray represent observations in one class (Class A) and the darker gray points represent observations in another class (Class B). The objective is to build a model

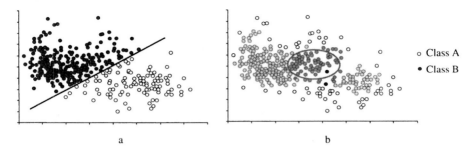

Figure 7.1. Classification of a categorical variable

that is able to classify observations into these two categories. In scatterplot a, a straight line can be drawn where observations above the line are placed in Class B and below the line observations are placed in Class A. In diagram b, observations in Class B (shown in dark gray) are grouped in the center of the scatterplot and observations in Class A are shown outside this central group. A model represented as an oval can be used to distinguish between these two groups. In many practical situations, the separation of observations between the different classes is not so simple. For example, Figure 7.2 illustrates how it is difficult to separate two classes.

The quality of a classification model can be assessed by counting the number of correctly and incorrectly assigned observations. For example, in the following contingency table the actual response is compared against the predicted response for a binary variable. The number of observations for each possible outcome are reported in Table 7.4.

This contingency table represents all possible outcomes for two binary variables:

- **Count:**$_{11}$ The number of observations that were true and predicted to be true (true positives).
- **Count:**$_{10}$ The number of observations that were false yet predicted to be true (false negatives).

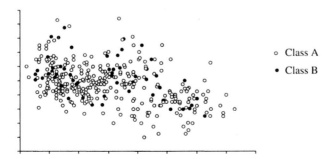

Figure 7.2. Categorical variables that would be difficult to classify based on the dimensions shown

Table 7.4. Contingency table showing a count of predicted vs actual values

		Predicted Response	
		True (1)	False (0)
Actual Response	True (1)	$Count_{11}$	$Count_{01}$
	False (0)	$Count_{10}$	$Count_{00}$

- **Count:**$_{01}$ The number of observations that were true and predicted to be false (false positives).
- **Count:**$_{00}$ The number of observations that were false and predicted to be false (true negatives).

In an ideal model there would be zero observation for $Count_{10}$ and $Count_{01}$. This is practically never the case, however, and the goal of any modeling exercise is to minimize the numbers for b and c according to criteria established in the definition step of the project. There are four calculations that are commonly used to assess the quality of a classification model:

- **Concordance:** This is an overall measure of the accuracy of the model and is calculated with the formula:

$$Concordance = \frac{(Count_{11} + Count_{00})}{(Count_{11} + Count_{10} + Count_{01} + Count_{00})}$$

- **Error rate:** This is an overall measure of the number of prediction errors and the formula is:

$$Error\ rate = \frac{(Count_{10} + Count_{01})}{(Count_{11} + Count_{10} + Count_{01} + Count_{00})}$$

- **Sensitivity:** This is an assessment of how well the model is able to predict 'true' values and the formula is:

$$Sensitivity = \frac{Count_{11}}{(Count_{11} + Count_{01})}$$

- **Specificity:** This is an assessment of how well the model is able to predict 'false' values and the formula is:

$$Specificity = \frac{Count_{00}}{(Count_{10} + Count_{00})}$$

For example, Table 7.5 shows the actual response values alongside the predicted response values for 18 observations. The contingency table is calculated from the actual and the predicted response values (Table 7.6). Based on this table, the

Table 7.5. Table showing an example of actual
and predicted values

Actual response	Predicted response
True (1)	True (1)
False (0)	False (0)
False (0)	False (0)
True (1)	True (1)
True (1)	False (0)
False (0)	True (1)
True (1)	True (1)
False (0)	False (0)
True (1)	True (1)
False (0)	False (0)
False (0)	False (0)
True (1)	True (1)
True (1)	False (0)
True (1)	True (1)
False (0)	False (0)
False (0)	False (0)
True (1)	True (1)
True (1)	True (1)

following assessments can be made of the accuracy of the model:

$$Concordance = \frac{(Count_{11} + Count_{00})}{(Count_{11} + Count_{10} + Count_{01} + Count_{00})}$$

$$= \frac{(8 + 7)}{(8 + 1 + 2 + 7)} = 0.83$$

$$Error\ rate = \frac{(Count_{10} + Count_{01})}{(Count_{11} + Count_{10} + Count_{01} + Count_{00})}$$

$$= \frac{(1 + 2)}{(8 + 1 + 2 + 7)} = 0.17$$

$$Sensitivity = \frac{Count_{11}}{(Count_{11} + Count_{01})} = \frac{8}{(8 + 2)} = 0.8$$

$$Specificity = \frac{Count_{00}}{(Count_{10} + Count_{00})} = \frac{7}{(1 + 7)} = 0.88$$

Table 7.6. Contingency table summarizing correct and incorrect
predictions

		Predicted Response	
		True (1)	False (0)
Actual Response	True (1)	8 (Count$_{11}$)	2 (Count$_{01}$)
	False (0)	1 (Count$_{10}$)	7 (Count$_{00}$)

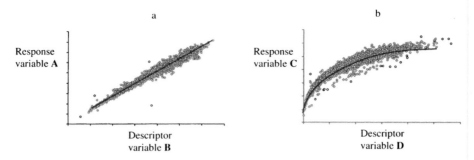

Figure 7.3. Linear and nonlinear relationships between continuous variables

In this example, the overall concordance is good with a 83% accuracy rate. The model is slightly better at predicting negatives than predicting positives and this is reflected in the higher specificity score.

The concordance gives an overall assessment of the accuracy of the model; however, based on the objectives of the modeling exercise it may be necessary to optimize on sensitivity or specificity if these are more important.

7.1.3 Regression

A regression model is a mathematical model that predicts a continuous response variable. For example, a regression model could be developed to predict actual sales volume or the temperature resulting from an experiment. Figure 7.3 shows two scatterplots. These illustrate the relationship between two variables. The variable on the y-axis is the response variable that is to be predicted. The variable on the x-axis is the descriptor variable that will be used in the predictive model. It is possible to see the relationship between the variables. In scatterplot a, as variable *B* increases, variable *A* increases proportionally. This relationship closely follows a straight line, as shown, and is called a *linear* relationship. In scatterplot b, as variable *D* increases, variable *C* also increases. In this case, the increase in *C* is not proportional to the increase in *D* and hence this type of relationship is called *nonlinear*. In Figure 7.4, it is not possible to see any relationship between the variables *F* and *E*.

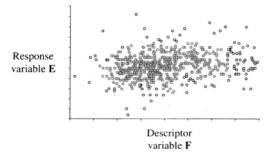

Figure 7.4. Scatterplot showing a difficult to discern relationship

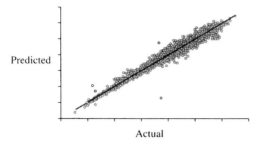

Figure 7.5. Scatterplot showing the results of a good prediction model

When the response value is continuous, one of the most informative ways of looking at the relationship between the actual values and the predicted values is a scatterplot, as shown in Figure 7.5. In the example in Figure 7.5, a line is drawn to indicate where the points would lie if the predicted values exactly matched the actual values, that is, the model was perfect. Good models have points lying close to this line. In Figure 7.6, the relationship between the actual response variable and the predicted value is shown. A line is drawn showing where the points should be placed if the prediction was perfect. In this situation, the model generated is poor since the actual predictions are scattered far from the line.

It is typical to use r^2 (described in Section 5.4.3) to describe the quality of the relationship between the actual response variable and the predicted response variable. Values for r^2 range between 0 and 1, with values closer to 1 indicating a better fit. Figure 7.7 shows two scatterplots displaying the relationship between predicted and actual response variables. The first scatterplot has an r^2 of 0.97, as the predicted values are close to the actual values, whereas the second scatterplot has an r^2 value of 0.07 since the model is poor.

The *residual* value is the difference between the actual value (y) and the predicted value ($\hat{y}$). Although the r^2 value provides a useful indication of the accuracy of the model, it is also important to look closely at the *residual* values.

$$residual = y - \hat{y}$$

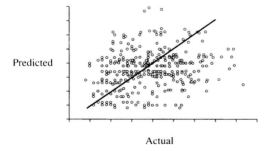

Figure 7.6. Scatterplot showing the results of a poor prediction model

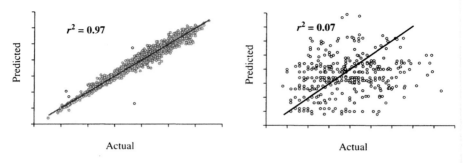

Figure 7.7. Calculated r^2 values for two prediction results

In Table 7.7, a residual value has been calculated using the actual (y) and the predicted ($\hat{y}$) values. It is important to analyze residuals based on a number of factors, including the following:

- **Response variable:** There should be no trends in residual values over the range of the response variable, that is, the distribution should be random.
- **Frequency distribution:** The frequency distribution of the residual values should follow a normal distribution.
- **Observation order:** There should be no discernable trends based on when the observations were measured.

Figure 7.8 illustrates an analysis of residuals for a simple model. The model is excellent as indicated by an r^2 value of 0.98. The scatterplot showing the residuals plotted against the response variable shows a reasonably even distribution of the residuals. The frequency distribution of the residual values follows a normal distribution. No trend can be seen in the residual values based on the order of the observations.

Table 7.7. Table showing example actual, predicted, and residual values

Actual response (y)	Predicted response ($\hat{y}$)	Residual
15.8	13.4	2.4
12.4	11.2	1.2
13.9	15.1	−1.2
8.4	8.3	0.1
6.6	5.2	1.4
16.4	16.9	−0.5

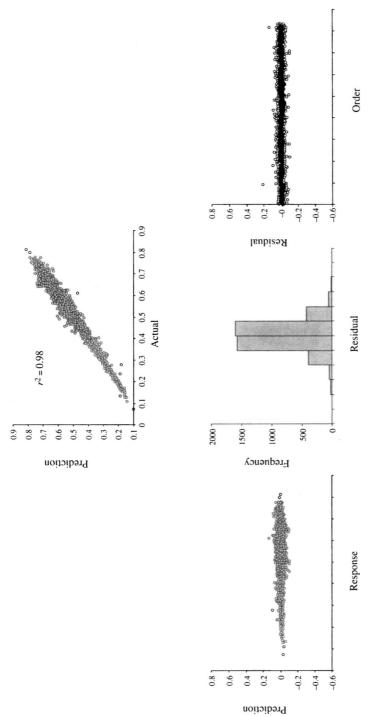

Figure 7.8. Analysis of residuals by response, frequency distribution, and order

7.1.4   Building a Prediction Model

Preparing the Data Set

It is important to prepare a data set prior to modeling as described in Chapter 3. This preparation should include the operations outlined such as characterizing, cleaning, and transforming the data. Particular care should be taken to determine whether subsetting the data is needed to simplify the resulting models.

Designing a Modeling Experiment

Building a prediction model is an experiment. It will be necessary to build many models for which you do not necessarily know which model will be the 'best'. This experiment should be appropriately designed to ensure an optimal result. There are three major dimensions that should be explored:

- **Different models:** There are many different approaches to building prediction models. A series of alternative models should be explored since all models work well in different situations. The initial list of modeling techniques to be explored can be based on the criteria previously defined as important to the project.

- **Different descriptor combinations:** Models that are based on a single descriptor are called *simple* models, whereas those built using a number of descriptors are called *multiple* (or multivariate) models. Correlation analysis as well as other statistical approaches can be used to identify which descriptor variables appear to be influential. A subject matter expert or business analyst may also provide insight into which descriptors would work best within a model. Care should be taken, however, not to remove variables too prematurely since interaction between variables can be significant within a model. Systematically trying different descriptor combinations to see which gives the best results can also be useful. In general, it is better to have fewer descriptors than observations.

- **Model parameters:** Most predictive models can be optimized by fine tuning different model parameters. Building a series of models with different parameter settings and comparing the quality of each model will allow you to optimize the model. For example, when building a neural network model there are a number of settings, which will influence the quality of the models built such as the number of cycles or the number of hidden layers. These parameters will be described in detail in Section 7.5.7.

Evaluating the 'best' model depends on the objective of the modeling process defined at the start of the project. Other issues, for example, the ability to explain how a prediction was made, may also be important and should be taken into account when assessing the models generated. Wherever possible, when two or more models give comparable results, the simpler model should be selected.

This concept of selecting the simplest model is often referred to as *Occam's Razor.*

Separating Test and Training Sets

The goal of building a predictive model is to generalize the relationship between the input descriptors and the output responses. The quality of the model depends on how well the model is able to predict correctly for a given set of input descriptors. If the model generalizes the input/output relationships too much, the accuracy of the model will be low. If the model does not generalize the relationships enough, then the model will have difficulties making predictions for observations not included in the data set used to build the model. Hence, when assessing the quality of the model, it is important to use a data set to build the model, which is different from the data set used to test the accuracy of the model. There are a number of ways for achieving this separation of test and training set, including the following:

- **Holdout:** At the start, the data set is divided up into a test and a training set. For example, a random 25% of the data set is assigned to the test set and the remaining 75% is assigned to the training set. The training set will be used to build the model and the test set will be used to assess the accuracy of the model.

- **Cross validation:** With cross validation methods, all observations in the data set will be used for testing and training, but not at the same time. Every observation will be assigned a predicted value and the difference between the predicted and the actual responses for all observations will be used to assess the model quality. To achieve this, it is necessary to assign a cross validation percentage. This number is the percentage of the data set that should be set aside for the test set at any one time. This percentage determines the number of models that are built. For example, a 5% cross validation will mean that, for each model, 5% of the data set will be set aside for testing and the remaining 95% will be used to build the model. To ensure that every example in the data set has a predicted value, 20 models will need to be built. There will also be 20 test sets (the complement of each training set), with no overlapping example between the different test sets. A cross validation where every observation is a separate test set, with the remaining observations used to build the models, is called a *leave-one-out* cross-validation.

7.1.5 Applying a Prediction Model

Once a model has been built and verified, it can be used to make predictions. Along with the presentation of the prediction, there should be some indications of the confidence in this value. It may be important to also provide an explanation of how the result was derived. Where the model is based on a simple and understandable formula, then this information may suffice. However, in many modeling techniques, a rather complex and nonintuitive mathematical formula may have been used and hence it is important to think about how to explain results in this situation. One option is to present portions of the training data with similar observations. Another alternative approach is to identify patterns or trends of observations that relate to the applied data.

During the data preparation step of the process, the descriptors and/or the response variables may have been translated to facilitate analysis. Once a prediction has been made, the variables should be translated back into their original format prior to presenting the information to the end user. For example, the log of the variable **Weight** was taken in order to create a new variable **log(Weight)** since the original variable was not normally distributed. This variable was used as a response variable in a model. Before any results are presented to the end user, the **log(Weight)** response should be translated back to **Weight** by taking the inverse of the log and presenting the value using the original weight scale.

A data set may have been divided or segmented into a series of simpler data sets in the data preparation step. Different models were developed from each. When applying these models to new data, some criteria will need to be established as to which model the observation will be presented to. For example, a series of models predicting house prices in different locations such as coastal, downtown, and suburbs were built. When applying these models to a new data set, the observations should be applied only to the appropriate model. Where a new observation can be applied to more than one model, some method for consolidating these potentially conflicting results will need to be established. A popular choice is often a voting scheme where the majority wins or the mean response for continuous variables.

In addition to building different models based on different criteria, multiple models may be built using different methods. Each model will provide a prediction and from these individual model predictions a final prediction may be calculated that is some function of these individual values. Techniques referred to as *Bagging* and *Boosting* can be used to accomplish this and further information on these methods is described in the further reading section of this chapter.

Once a model has been built, a useful exercise is to look at the observations that were not correctly predicted. Attempting to understand any relationship within these observations can be important in understanding whether there is a problem with these observations. For example, if all incorrectly assigned observations were measured using a particular device, then perhaps there was a problem with the calibration of this measuring device. The observations may also share additional common characteristics. Understanding the grouping of these observations using techniques such as clustering may help to suggest why the model is unable to correctly predict these examples. It may suggest that additional descriptors are required in order to adequately predict this type of observation. It may also suggest that these types of observations should not be presented to the model in the future.

7.2 SIMPLE REGRESSION MODELS

7.2.1 Overview

A simple regression model is a formula describing the relationship between one descriptor variable and one response variable. These formulas are easy to explain; however, the analysis is sensitive to any outliers in the data. The following section presents methods for generating simple linear regression models as well as simple nonlinear regression models.

7.2.2 Simple Linear Regression

Overview

Where there appears to be a linear relationship between two variables, a simple linear regression model can be generated. For example, Figure 7.9 shows the relationship between a descriptor variable **B** and a response variable **A**. The diagram shows a high degree of correlation between the two variables. As descriptor variable **B** increases, response variable **A** increases at the same rate. A straight line representing a model can be drawn through the center of the points. A model that would predict values along this line would provide a good model.

A straight line can be described using the formula:

$$y = a + bx$$

where a is the point of intersection with the y-axis and b is the slope of the line. This is shown graphically in Figure 7.10.

In Table 7.8, a data set of observations from a grocery store contains variables **Income** and **Monthly sales**. The variable **Income** refers to the yearly income for a customer and the **Monthly sales** represent the amount that particular customer purchases per month. This data can be plotted on a scatterplot

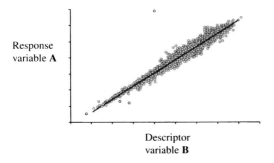

Figure 7.9. Scatterplot illustrating a simple linear model

Table 7.8.   Table of the customer's Income and Monthly Sales

Income (x)	Monthly Sales (y)
$15,000.00	$54.00
$16,000.00	$61.00
$17,000.00	$70.00
$18,000.00	$65.00
$19,000.00	$68.00
$20,000.00	$84.00
$23,000.00	$85.00
$26,000.00	$90.00
$29,000.00	$87.00
$33,000.00	$112.00
$35,000.00	$115.00
$36,000.00	$118.00
$38,000.00	$120.00
$39,000.00	$118.00
$41,000.00	$131.00
$43,000.00	$150.00
$44,000.00	$148.00
$46,000.00	$151.00
$49,000.00	$157.00
$52,000.00	$168.00
$54,000.00	$156.00
$52,000.00	$158.00
$55,000.00	$161.00
$59,000.00	$183.00
$62,000.00	$167.00
$65,000.00	$186.00
$66,000.00	$191.00

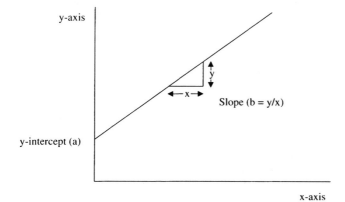

Figure 7.10.   Calculation of the slope of a straight line

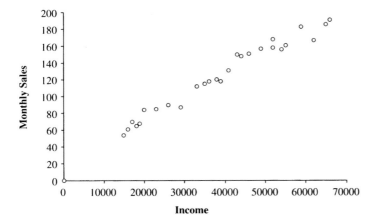

Figure 7.11. Scatterplot of the customer's **Income vs Monthly Sales**

and a linear relationship between **Income** and **Monthly sales** can be seen in Figure 7.11.

To manually generate a linear regression formula, a straight line is drawn through the points as shown in Figure 7.12. The point at which the line intercepts with the y-axis is noted (approximately 20) and the slope of the line is calculated (approximately 50/20,000 or 0.0025). For this data set an approximate formula for the relationship between **Income** and **Monthly sales** is:

$$\textbf{Monthly sales} = 20 + 0.0025 \times \textbf{Income}$$

Once a formula for the straight line has been established, predicting values for the y response variable based on the x descriptor variable can be easily calculated. The formula should only be used, however, for values of the x variable within the range from which the formula was derived. In this example, **Monthly sales** should only

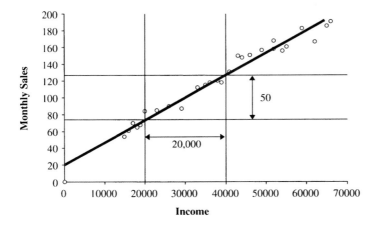

Figure 7.12. Calculating the slope of the line

be predicted based on **Income** between $15,000 and $66,000. A prediction for a customer's **Monthly sales** based on their **Income** can be calculated. For a customer with an **Income** of $31,000, the **Monthly sales** would be predicted as:

$$\textbf{Monthly sales} \ (\text{predicted}) = 20 + 0.0025 \times \$31,000$$
$$\textbf{Monthly sales} \ (\text{predicted}) = \$97.50$$

Least Squares Method

Parameters *a* and *b* can be derived manually by drawing a best guess line through the points in the scatterplot and then visually inspecting where the line crosses the y-axis (*a*) and measuring the slope (*b*) as previously described. The least squares method is able to calculate these parameters automatically. The formula for calculating a slope is:

$$b = \frac{\sum_{i=1}^{n}(x_i - \bar{x})(y_i - \bar{y})}{\sum_{i=1}^{n}(x_i - \bar{x})^2}$$

where x_i and y_i are the individual values for the descriptor variable (x_i) and the response (y_i). $\bar{x}$ is the mean of the descriptor variable x and $\bar{y}$ is the mean of the response variable y.

The formula for calculating the intercept with the y-axis is:

$$a = \bar{y} - b\bar{x}$$

Using the data from Table 7.8, the slope and intercept are calculated using Table 7.9. The average **Income** is $38,963 and the average **Monthly sales** is $124.22.

$$\text{Slope} \ (b) = 17,435,222/6,724,962,963$$
$$\text{Slope} \ (b) = 0.00259$$

$$\text{Intercept} \ (a) = 124.22 - (0.00259 \times 38,963)$$
$$\text{Intercept} \ (a) = 23.31$$

Hence the formula is:

$$\textbf{Monthly sales} = 23.31 + 0.00259 \times \textbf{Income}$$

These values are close to the values calculated using the manual approach.

Most statistical packages will calculate a simple linear regression formula automatically.

7.2.3   Simple Nonlinear Regression

In situations where the relationship between two variables is nonlinear, a simple way of generating a regression equation is to transform the nonlinear relationship to a

Table 7.9. Calculation of linear regression with least squares method

Income (x)	Monthly Sales (y)	$(x_i - \bar{x})$	$(y_i - \bar{y})$	$(x_i - \bar{x})$ $(y_i - \bar{y})$	$(x_i - \bar{x})^2$
$15,000.00	$54.00	−23,963	−70.22	1,682,733	574,223,594
$16,000.00	$61.00	−22,963	−63.22	1,451,770	527,297,668
$17,000.00	$70.00	−21,963	−54.22	11,908,801	482,371,742
$18,000.00	$65.00	−20,963	−59.22	1,241,473	439,445,816
$19,000.00	$68.00	−19,963	−56.22	1,122,362	398,519,890
$20,000.00	$84.00	−18,963	−40.22	762,733	359,593,964
$23,000.00	$85.00	−15,963	−39.22	626,103	254,816,187
$26,000.00	$90.00	−12,963	−34.22	443,621	168,038,409
$29,000.00	$87.00	−9,963	−37.22	370,844	99,260,631
$33,000.00	$112.00	−5,963	−12.22	72,881	35,556,927
$35,000.00	$115.00	−3,963	−9.22	36,547	15,705,075
$36,000.00	$118.00	−2,963	−6.22	18,436	8,779,150
$38,000.00	$120.00	−963	−4.22	4,066	927,298
$39,000.00	$118.00	37	−6.22	−230	1,372
$41,000.00	$131.00	2,037	6.78	13,807	4,149,520
$43,000.00	$150.00	4,037	25.78	104,066	16,297,668
$44,000.00	$148.00	5,037	23.78	119,770	25,371,742
$46,000.00	$151.00	7,037	26.78	188,436	49,519,890
$49,000.00	$157.00	10,037	32.78	328,992	100,742,112
$52,000.00	$168.00	13,037	43.78	570,733	169,964,335
$54,000.00	$156.00	15,037	31.78	477,844	226,112,483
$52,000.00	$158.00	13,037	33.78	440,362	169,964,335
$55,000.00	$161.00	16,037	36.78	589,807	257,186,557
$59,000.00	$183.00	20,037	58.78	1,177,733	401,482,853
$62,000.00	$167.00	23,037	42.78	985,473	530,705,075
$65,000.00	$186.00	26,037	61.78	1,608,510	677,927,298
$66,000.00	$191.00	27,037	66.78	1,805,473	731,001,372
Totals				17,435,222	6,724,962,963

linear relationship using a mathematical transformation. A linear model (as described above) can then be generated. Once a prediction has been made, the predicted value is transformed back to the original scale. For example, in Table 7.10 two columns show a nonlinear relationship. Plotting these values results in the scatterplot in Figure 7.13.

There is no linear relationship between these two variables and hence we cannot calculate a linear model directly from the two variables. To generate a model, we transform *x* or *y* or both to create a linear relationship. In this example, we transform the *y* variable using the following formula:

$$y' = \frac{-1}{y}$$

Table 7.10. Table of observations
for variables x and y

x	y
3	4
6	5
9	7
8	6
10	8
11	10
12	12
13	14
13.5	16
14	18
14.5	22
15	28
15.2	35
15.3	42

We now generate a new column, y' (Table 7.11). If we now plot x against y', we can see that we now have an approximate linear relationship (see Figure 7.14). Using the least squares method described previously, an equation for the linear relationship between x and y' can be calculated. The equation is:

$$y' = -0.307 + 0.018 \times x$$

Using x we can now calculate a predicted value for the transformed value of y (y'). To map this new prediction of y' we must now perform an inverse transformation, that is, $-1/y'$. In Table 7.12, we have calculated the predicted value for y' and

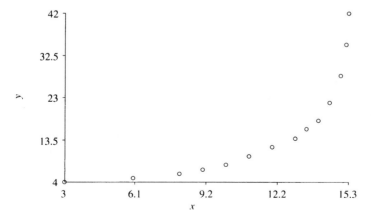

Figure 7.13. Scatterplot showing the nonlinear relationship between x and y

Table 7.11. Transformation of y to create a linear relationship

X	y	$y' = -1/y$
3	4	−0.25
6	5	−0.2
9	7	−0.14286
8	6	−0.16667
10	8	−0.125
11	10	−0.1
12	12	−0.08333
13	14	−0.07143
13.5	16	−0.0625
14	18	−0.05556
14.5	22	−0.04545
15	28	−0.03571
15.2	35	−0.02857
15.3	42	−0.02381

transformed the number to **Predicted y**. The **Predicted y** values are close to the actual **y** values.

Some common nonlinear relationships are shown in Figure 7.15. The following transformation may create a linear relationship for the charts shown:

- **Situation a:** Transformations on the x, y or both x and y variables such as *log* or *square root*.

- **Situation b:** Transformation on the x variable such as *square root*, *log* or $-1/x$.

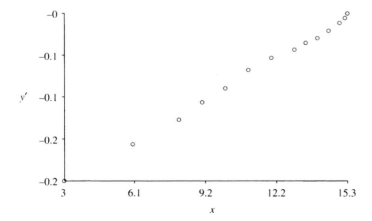

Figure 7.14. Scatterplot illustrating the new linear relationship

Table 7.12.   Prediction of y using a nonlinear model

x	y	$y' = -1/y$	Predicted y'	Predicted y
3	4	−0.25	−0.252	3.96
6	5	−0.2	−0.198	5.06
9	7	−0.143	−0.143	6.99
8	6	−0.167	−0.161	6.20
10	8	−0.125	−0.125	8.02
11	10	−0.1	−0.107	9.39
12	12	−0.083	−0.088	11.33
13	14	−0.071	−0.070	14.28
13.5	16	−0.062	−0.061	16.42
14	18	−0.056	−0.052	19.31
14.5	22	−0.045	−0.043	23.44
15	28	−0.036	−0.033	29.81
15.2	35	−0.029	−0.023	33.45
15.3	42	−0.024	−0.028	35.63

- **Situation c:** Transformation on the y variable such as *square root, log* or $-1/y$.

This approach to creating simple nonlinear models can only be used when there is a clear transformation of the data to a linear relationship. Other methods described later in this chapter can be used where this is not the case.

7.3   K-NEAREST NEIGHBORS

7.3.1   Overview

The k-Nearest Neighbors (kNN) method provides a simple approach to calculating predictions for unknown observations. It calculates a prediction by looking at similar observations and uses some function of their response values to make the prediction, such as an average. Like all prediction methods, it starts with a training set but instead of producing a mathematical model it determines the optimal number of similar observations to use in making the prediction.

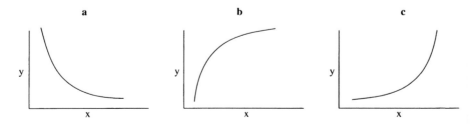

Figure 7.15.   Nonlinear scenarios

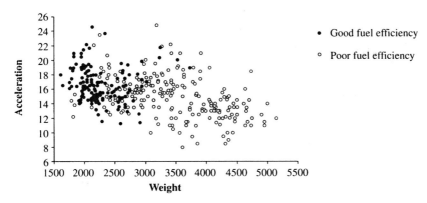

Figure 7.16. Scatterplot showing fuel efficiency classifications

    The scatterplot in Figure 7.16 is based on a data set of cars and will be used to illustrate how kNN operates. Two variables that will be used as descriptors are plotted on the *x*- and *y*-axis (**Weight** and **Acceleration**). The response variable is a dichotomous variable (**Fuel Efficiency**) that has two values: good and poor fuel efficiency. The darker shaded observations have good fuel efficiency and the lighter shaded observations have poor fuel efficiency.

    During the learning phase, the best number of similar observations is chosen (*k*). The selection of *k* is described in the next section. Once a value for *k* has been determined, it is now possible to make a prediction for a car with unknown fuel efficiency. To illustrate, two cars with unknown fuel efficiency are presented to the kNN model in Figure 7.17: A and B. The **Acceleration** and **Weight** of these observations are known and the two observations are plotted alongside the training set. Based on the optimal value for *k*, the *k* most similar observations to A and B are identified in Figure 7.18. For example, if *k* was calculated to be 10, then the 10 most similar observations from the training set would be selected. A prediction is made

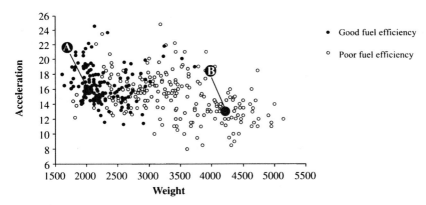

Figure 7.17. Two new observations (A and B) plotted alongside existing data

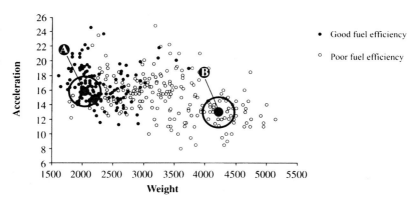

Figure 7.18. Looking at similar observations to predict values for A and B

for A and B based on the response of the nearest neighbors. In this case, observation A would be predicted to have good fuel efficiency since its neighbors all have good fuel efficiency. Observation B would be predicted to have poor fuel efficiency since its neighbors all have poor fuel efficiency.

kNN has a number of advantages:

- **Noise:** kNN is relatively insensitive to errors or outliers in the data.
- **Large sets:** kNN can be used with large training sets.

kNN has the following disadvantage:

- **Speed:** kNN can be computationally slow when it is applied to a new data set since a similar score must be generated between the observations presented to the model and every member of the training set.

7.3.2 Learning

A kNN model uses the k most similar neighbors to the observation to calculate a prediction. Where a response variable is continuous, the prediction is the mean of the nearest neighbors. Where a response variable is categorical, the prediction could be presented as a mean or a voting scheme could be used, that is, select the most common classification term.

In the learning phase, three items should be determined:

- **Best similarity method:** As described in Chapter 6, there are many methods for determining whether two observations are similar. For example, the Euclidean or the Jaccard distance. Prior to calculating the similarity, it is important to normalize the variables to a common range so that no variables are considered to be more important.
- k: This is the number of similar observations that produces the best predictions. If this value is too high, then the kNN model will overgeneralize. If the value is too small, it will lead to a large variation in the prediction.

- **Combination of descriptors:** It is important to understand which combination of descriptors results in the best predictions.

The selection of k is performed by adjusting the values of k within a range and selecting the value that gives the best prediction. To ensure that models generated using different values of k are not overfitting, a separate training and test set should be used.

To assess the different values for k, the sum of squares of error (SSE) evaluation criteria will be used:

$$SSE = \sum_{i=1}^{k} (\hat{y}_i - \bar{y})^2$$

Smaller SSE values indicate that the predictions are closer to the actual values. To illustrate, a data set of cars will be used and a model built to test the car fuel efficiency (**MPG**). The following variables will be used as descriptors within the model: **Cylinders, Displacement, Horsepower, Weight, Acceleration, Model Year** and **Origin**. The Euclidean distance calculation was selected to represent the distance between observations. To calculate an optimal value for k, different values of k were selected between 2 and 20. To test the models built with the different values of k, a 10% cross-validation split was made to ensure that the models were built and tested with different observations. The SSE evaluation criterion was used to assess the quality of each model. In this example, the value of k with the lowest SSE value is 6 and this value is selected for use with the kNN model (see Table 7.13).

Table 7.13. Table for detecting the best values for k

K	SSE
2	3,533
3	3,414
4	3,465
5	3,297
6	3,218
7	3,355
8	3,383
9	3,445
10	3,577
11	3,653
12	3,772
13	3,827
14	3,906
15	3,940
16	3,976
17	4,058
18	4,175
19	4,239
20	4,280

Table 7.14. Observation to be predicted

Names	Cyclinders	Displace- ment	Horse- power	Weight	Accele- ration	Model Year	Origin
Dodge Aspen	6	225	90	3,381	18.7	1980	1

7.3.3 Predicting

Once a value for k has been set in the training phase, the model can now be used to make predictions. For example, an observation x has values for the descriptor variables but not for the response. Using the same technique for determining similarity as used in the model building phase, observation x is compared against all observations in the training set. A distance is computed between x and each training set observation. The closest k observations are selected and a prediction is made, for example, using the average value.

The observation (Dodge Aspen) in Table 7.14 was presented to the kNN model built to predict car fuel efficiency (**MPG**). The Dodge Aspen observation was compared to all observations in the training set and an Euclidean distance was computed. The six observations with the smallest distance scores are selected, as shown in Table 7.15. The prediction is the average of these top six observations, that is, 19.5. In Table 7.16, the cross validated prediction is shown alongside the actual value.

Table 7.15. Table of similar observations

Names	Cyclinders	Displace- ment	Horse- power	Weight	Accele- ration	Model Year	Origin	MPG
Chrysler Lebaron Salon	6	225	85	3,465	16.6	1981	1	17.6
Mercury Zephyr 6	6	200	85	2,990	18.2	1979	1	19.8
Ford Granada GL	6	200	88	3,060	17.1	1981	1	20.2
Pontiac Phoenix LJ	6	231	105	3,535	19.2	1978	1	19.2
AMC Concord	6	232	90	3,210	17.2	1978	1	19.4
Plymouth Volare	6	225	100	3,430	17.2	1978	1	20.5
							Average	19.5

Table 7.16. Actual vs predicted values

Names	Cyclinders	Displace-ment	Horse-power	Weight	Accele-ration	Model Year	Origin	Actual MPG	Predicted MPG
Dodge Aspen	6	225	90	3,381	18.7	1980	1	19.1	19.5

The training set of observations can be used to explain how the prediction was reached in addition to assessing the confidence in the prediction. For example, if the response values for these observations were all close this would increase the confidence in the prediction.

7.4 CLASSIFICATION AND REGRESSION TREES

7.4.1 Overview

In Chapter 6, decision trees were described as a way of grouping observations based on specific values or ranges of descriptor variables. For example, the tree in Figure 7.19 organizes a set of observations based on the number of cylinders (**Cylinders**) of the car. The tree was constructed using the variable **MPG** (miles per gallon) as the response variable. This variable was used to guide how the

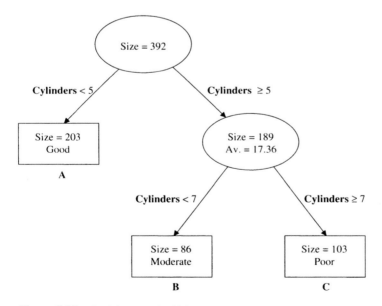

Figure 7.19. Decision tree classifying cars

tree was constructed, resulting in groupings that characterize car fuel efficiency. The terminal nodes of the tree (A, B, and C) show a partitioning of cars into sets with good (node A), moderate (node B), and poor (node C) fuel efficiencies.

Each terminal node is a mutually exclusive set of observations, that is, there is no overlap between nodes A, B, or C. The criteria for inclusion in each of these nodes are defined by the set of branch points used to partition the data. For example, terminal node B is defined as observations where **Cylinders** are greater or equal to five and **Cylinders** are less than seven.

Decision trees can be used as both classification and regression prediction models. Decision trees that are built to predict a continuous response variable are called *regression trees* and decision trees built to predict a categorical response are called *classification trees*. During the learning phase, a decision tree is constructed as before using the training set. Predictions in decision trees are made using the criteria associated with the terminal nodes. A new observation is assigned to a terminal node in the tree using these splitting criteria. The prediction for the new observation is either the node classification (in the case of a classification tree) or the average value (in the case of a regression tree). In the same way as other prediction modeling approaches, the quality of the prediction can be assessed using a separate training set.

7.4.2 Predicting Using Decision Trees

In Figure 7.20, a set of cars is shown on a scatterplot. The cars are defined as having good fuel efficiency or poor fuel efficiency. Those with good fuel efficiency are shaded darker than those with poor fuel efficiency. Values for the **Acceleration** and **Weight** variables are shown on the two axes.

A decision tree is generated using the car fuel efficiency as a response variable. This results in a decision tree where the terminal nodes partition the set of

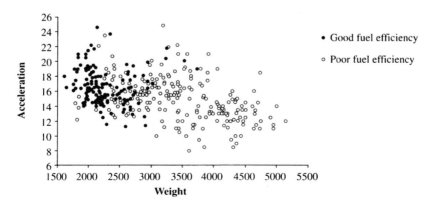

Figure 7.20. Distribution of cars classified by fuel efficiency

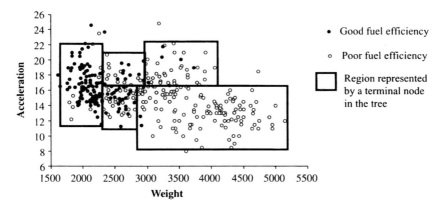

Figure 7.21. Dividing cars into regions based on classifications

observations according to ranges in the descriptor variables. One potential partition of the data is shown in Figure 7.21. The prediction is then made based on the observations used to train the model that are within the specific region, such as the most popular class or the average value (see Figure 7.22).

When an observation with unknown fuel efficiency is presented to the decision tree model, it is placed within one of the regions. The placement is based on the observation's descriptor values. Two observations (A and B) with values for **Acceleration** and **Weight**, but no value for whether the cars have good or poor fuel efficiency, are presented to the model. These observations are shown on the scatterplot in Figure 7.23. Observation A will be predicted to have good fuel efficiency whereas observation B will be predicted to have poor fuel efficiency.

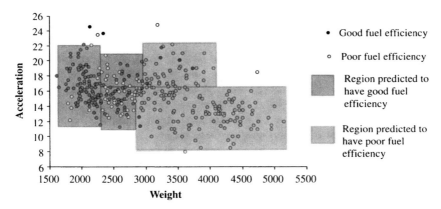

Figure 7.22. Assigning prediction categories to regions

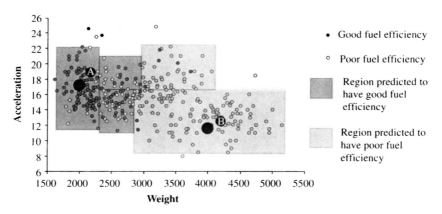

Figure 7.23. Prediction of two unknown observations

Decision trees are useful for prediction since the results are easy to explain. Unfortunately, these types of models can be quite sensitive to any noise in the training set.

The same parameters used to build the tree (described in Section 6.4) can be set to build a decision tree model, that is, different input descriptor combinations and different stopping criteria for the tree.

7.4.3 Example

The decision tree in Figure 7.24 was built from a data set of 352 cars, using the continuous variable **MPG** to split the observations. The average value shown in the diagram is the **MPG** value for the set of observations. The nodes were not split further if there were less than 30 observations in the terminal node.

In Table 7.17, a set of 36 observations not used in building the tree are shown with both an actual and a predicted value. The final column indicates the node in the tree that was used to calculate the prediction. For example, the AMC Gremlin with a **Horsepower** of 90 and **Weight** of 2648 will fit into a region defined by node D in the tree. Node D has an average **MPG** value of 23.96 and hence this is the predicted **MPG** value. The table also indicates the actual **MPG** values for the cars tested.

The examples used in this section were simple in order to describe how predictions can be made using decision trees. It is usual to use larger numbers of descriptor variables. Also, building a series of models based on changing the terminating criteria can also be useful in optimizing the decision tree models. The further reading section of chapter 6 provides references to additional methods for optimizing decision trees.

The terminal nodes in the decision trees can be described as rules, (as shown in Section 6.3) which can be useful in explaining how a prediction was obtained. In addition, looking at the data that each rule is based on allows us to understand the degree of confidence with which each prediction was made. For example, the

c07f024.eps

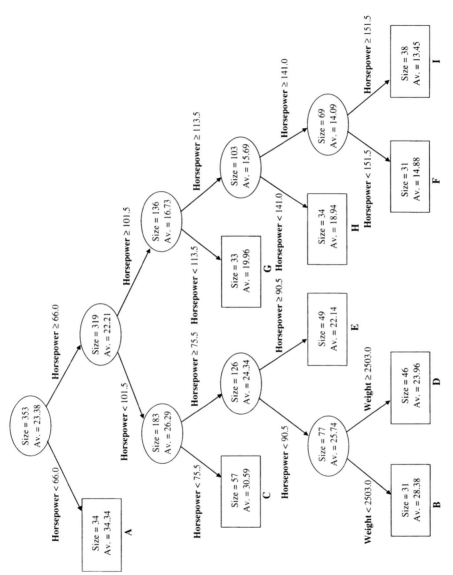

Figure 7.24. Decision tree generated for prediction

185

Table 7.17. Predictions based on terminal node averages

Names	Horsepower	Weight	MPG (Actual)	MPG (Predicted)	Rule ID
AMC Gremlin	90	2,648	21	23.96	D
AMC Matador	120	3,962	15.5	18.94	H
AMC Rebel SST	150	3,433	16	14.88	F
AMC Spirit DL	80	2,670	27.4	23.96	D
BMW 2002	113	2,234	26	19.96	G
Buick Century Limited	110	2,945	25	19.96	G
Buick Skylark	84	2,635	26.6	23.96	D
Chevrolet Chevette	63	2,051	30.5	34.34	A
Chevrolet Impala	165	4,274	13	13.45	I
Chevrolet Monza 2 + 2	110	3,221	20	19.96	G
Chevrolet Nova	100	3,336	15	22.14	E
Chrysler Lebaron Medallion	92	2,585	26	22.14	E
Datsun 310 GX	67	1,995	38	30.59	C
Datsun b210	67	1,950	31	30.59	C
Dodge Aries Wagon (SW)	92	2,620	25.8	22.14	E
Dodge Aspen	110	3,620	18.6	19.96	G
Dodge Colt Hatchback Custom	80	1,915	35.7	28.38	B
Fiat 124 TC	75	2,246	26	30.59	C
Ford Fairmont (man)	88	2,720	25.1	23.96	D
Ford Fiesta	66	1,800	36.1	30.59	C
Ford Gran Torino	152	4,215	14.5	13.45	I
Ford Mustang II 2 + 2	89	2,755	25.5	23.96	D
Ford Pinto	80	2,451	26	28.38	B
Ford Pinto Runabout	86	2,226	21	28.38	B
Honda Accord LX	68	2,135	29.5	30.59	C
Maxda GLC Deluxe	65	1,975	34.1	34.34	A
Mercury Marquis Brougham	198	4,952	12	13.45	I
Nissan Stanza XE	88	2,160	36	28.38	B
Plymouth Reliant	84	2,490	27.2	28.38	B
Plymouth Valiant	100	3,233	22	22.14	E
Plymouth Volare	100	3,430	20.5	22.14	E
Pontiac Catalina	175	4,385	14	13.45	I
Pontiac Safari (SW)	175	5,140	13	13.45	I
Toyota Corona Mark II	95	2,372	24	22.14	E
Toyota Tercel	62	2,050	37.7	34.34	A
Volvo 245	102	3,150	20	19.96	G

number of observations and the distribution of the response variable can help to understand how much confidence we should have in the prediction.

7.5 NEURAL NETWORKS

7.5.1 Overview

A neural network is a mathematical model that makes predictions based on a series of input descriptor variables. Like all prediction models, it uses a training set of examples to generate the model. This training set is used to generalize the relationships between the input descriptor variables and the output response variables. Once a neural network has been created, it can then be used to make predictions. The following sections describe what neural networks look like, how they learn and how they make predictions. An example is presented illustrating how neural networks can be optimized.

7.5.2 Neural Network Layers

A neural network comprises of a series of independent processors or nodes. These nodes are connected to other nodes and are organized into a series of layers as shown in Figure 7.25. In this example, each node is assigned a letter from A to L and organized into three layers. The input layer contains a set of nodes (A, B, C, D, E, F). Each node in the input layer corresponds to a numeric input descriptor variable. In this case, there are six input descriptor variables. The layer shown in black is the output layer containing nodes K and L. Each output node corresponds to an output response variable (two in this example). Between the input layer and the output layer is a hidden layer of nodes (G, H, I, J). In this example, there is just a single hidden layer comprised of four nodes. The number of hidden layers normally range from 0

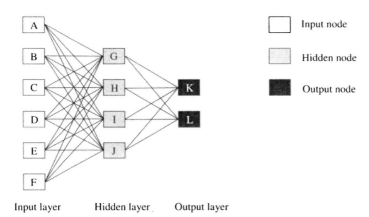

Figure 7.25. Topology of a neural network

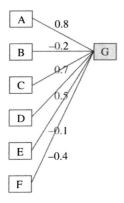

Figure 7.26. Weights associated with each connection

to 5. Each node in the network is often connected to all nodes in the layers adjacent to the node. For example, node G is connected to A, B, C, D, E, F in the input layer and nodes K and L in the output layer.

Every connection, such as between A and G, has a number or *weight* associated with it. Prior to learning, the weights are assigned random values usually in the range -1 to $+1$. These weights will be adjusted during the learning process. In Figure 7.26, a portion of the neural network is displayed, showing node G along with the nodes connected to it. Random numbers between -1 and $+1$ are assigned to each connection. For example, the connection between A and G is randomly assigned a weight of 0.8.

7.5.3 Node Calculations

Each node in the neural network calculates a single output value based on a set of input values (I_1 to I_n), as shown in Figure 7.27. For nodes in the first hidden layer, the input values correspond to the input descriptor values.

Each input connection has a weight and is assigned a value. The total input of the node is calculated using these weights and values. For example, the following formula for calculating the combined input is often used:

$$Input = \sum_{j=1}^{n} I_j w_j$$

where I_j are the individual input values and w_j are the individual weights.

In this example, the observation in Table 7.18 is presented to the network with values normalized between 0 and 1. The observation has six descriptor variables and two response variables. The six descriptor variables are labeled V_1 to V_6 and the two response variables are labeled V_7 and V_8. The input descriptor variables are presented to the neural network, as shown in Figure 7.28. V_1 is presented to node A, V_2 to node B, etc. These inputs all feed into node G. The

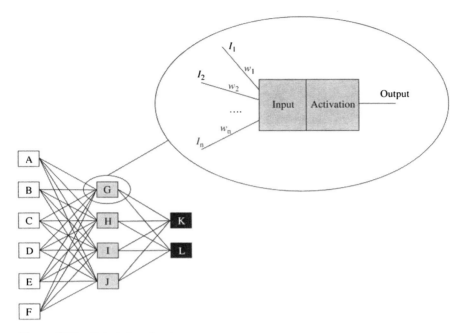

Figure 7.27.   Calculation of node output

combined input value for node G is calculated using the input values and the weights:

$$Input_G = \sum_{j=1}^{n} I_j w_j$$

$$Input_G = (1 \times 0.8) + (0 \times -0.2) + (1 \times 0.7) + (1 \times 0.5)$$
$$+ (0.5 \times -0.1) + (0.8 \times -0.4)$$

$$Input_G = 1.63$$

For a number of reasons, this combined input value is now processed further using an *activation function*. This function will generate the output for the node. Common activation functions include:

$$\textbf{Sigmoid}: \quad Output = \frac{1}{1 + e^{-Input}}$$

$$\textbf{Tanh}: \quad Output = \frac{e^{Input} - e^{-Input}}{e^{Input} + e^{-Input}}$$

Table 7.18.   Example observation with six inputs (descriptors) and two outputs (responses)

Descriptor variables						Response variables	
V_1	V_2	V_3	V_4	V_5	V_6	V_7	V_8
1	0	1	1	0.5	0.8	0.4	1

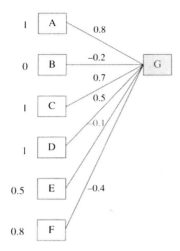

Figure 7.28.   Presenting input values to the neural network

These types of activation functions allow the neural network to develop nonlinear models. The *sigmoid* function will produce an output between 0 and +1 and the *tanh* function will produce an output between −1 and +1.

Using the above example with the *sigmoid* activation function, the following output from the neural network node G would be generated:

$$Output_G = \frac{1}{1 + e^{-Input_G}}$$

$$Output_G = \frac{1}{1 + e^{-1.63}}$$

$$Output_G = 0.84$$

7.5.4   Neural Network Predictions

A neural network makes a prediction based on the input descriptor variables presented to the network and the weights associated with connections in the network. For example, Figure 7.29 shows an observation presented to the network.

The first hidden layer uses these input values along with the weights associated with the connections between the input nodes and the hidden layer nodes to calculate an output (as described previously). Each of these outputs is then presented to the nodes in the next layer. These values are now inputs to the next layer. In this neural network, there is only one hidden layer and so the outputs from nodes G, H, I, J are now the inputs to nodes K and L. These input values are combined with the weights of the connections. Nodes K and L each produce a single output corresponding to the two response variables. These values are the predictions. The process of taking a set

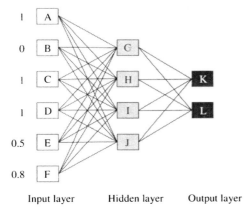

Figure 7.29.   Input variables presented to network

of input descriptors and calculating one or more output responses is called *feed forward.*

Initially, all the weights in the neural network are randomly assigned and hence these initial predictions will be meaningless. These weights will be adjusted during the learning process resulting in predictions with greater predictive accuracy.

7.5.5 Learning Process

All input/output values in the training set should be normalized prior to training the network. This is to avoid introducing unnecessary bias resulting from variables being measured on different scales. The learning process proceeds by taking random examples from the training set, which are then presented to the neural network. The neural network then makes a prediction. The neural network will learn by adjusting the weights according to how well the predictions match the actual response values. The observation from Table 7.18 is presented to the neural network and the network calculates predictions for the two response variables. Node K generates a prediction for response variable V_7 and node L generates a prediction for response variable V_8, as shown in Figure 7.30.

In this example, the neural network has not started to learn and hence the predictions are not close to the actual response values. The *error* or difference between the actual responses and the predicted responses is calculated. The neural network then attempts to learn by adjusting the weights of the network using this error. Once the weights have been adjusted, the network is presented with another random example from the training set and the weights are again adjusted based on this new example. As more and more examples are presented to the network, the error between the predicted responses and the actual responses gets smaller. At a certain point, the learning process stops and the network is ready to be used for prediction. Figure 7.31 illustrates the learning process.

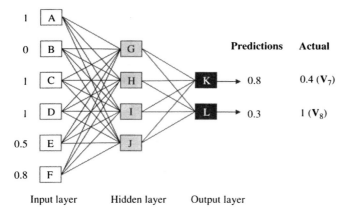

Input layer   Hidden layer   Output layer

Figure 7.30.   Comparing predicted against actual values

7.5.6 Backpropagation

One of the most commonly used techniques for learning in neural networks is called *backpropagation*. In order for the weights of the neural network connections to be adjusted, an error first needs to be calculated between the predicted response and the actual response. The following formula is commonly used for the output layer:

$$Error_i = Output_i(1 - Output_i)(Actual_i - Output_i)$$

where $Error_i$ is the error resulting from node i, $Output_i$ is the predicted response value and $Actual_i$ is the actual response value.

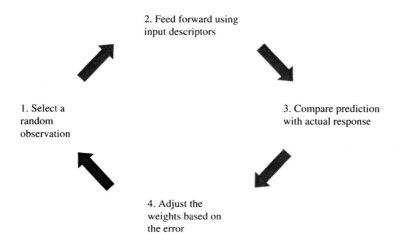

Figure 7.31.   Learning process in neural networks

For example, the errors calculated for nodes K and L are:

Node K:

$Error_K = Output_K(1 - Output_K)(Actual_K - Output_K)$

$Error_K = 0.8 \times (1 - 0.8) \times (0.4 - 0.8)$

$Error_K = -0.064$

Node L:

$Error_L = Output_L(1 - Output_L)(Actual_L - Output_L)$

$Error_L = 0.3 \times (1 - 0.3) \times (1 - 0.3)$

$Error_L = 0.147$

Once the error has been calculated for the output layer, it can now be backpropagated, that is, the error can be passed back through the neural network. To calculate an error value for the hidden layers, the following calculation is commonly used:

$$Error_i = Output_i(1 - Output_i) \sum_{j=1}^{n} Error_j w_{ij}$$

where $Error_i$ is the error resulting from the hidden node, $Output_i$ is the value of the output from the hidden node, $Error_j$ is the error already calculated for the jth node connected to the output and w_{ij} is the weight on this connection.

Figure 7.32 illustrates how the errors are calculated for nodes other than the output layer. After calculating the error for nodes K and L, the error of the hidden layer can be calculated. Node G is used as an example for a hidden layer error calculation as shown below.

Node G:

$Error_G = Output_G(1 - Output_G)((Error_K \times w_{GK}) + (Error_L \times w_{GL}))$

$Error_G = 0.84 \times (1 - 0.84) \times ((-0.064 \times 0.3) + (0.147 \times -0.7))$

$Error_G = 0.0112$

An error should be calculated for all output and hidden layer nodes. Errors for hidden layer nodes use errors from the nodes their output is attached to, which have

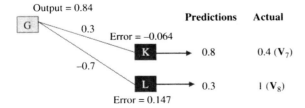

Figure 7.32.   Comparing errors in the output layer

already been calculated. Once the error has been propagated throughout the neural network, the error values can be used to adjust the weights of the connections using the formula:

$$w_{ij} = w_{ij} + l \times Error_j \times Output_i$$

where w_{ij} is the weight of the connection between nodes i and j, $Error_j$ is the calculated error for node j, $Output_i$ is the computed output from node i, and l is the predefined learning rate. This takes a value between 0 and 1. The smaller the learning rate value, the slower the learning process. Often a learning rate is set high initially and then reduced as the network fine-tunes the weights.

To calculate the new weight for the connection between G and K where the learning rate (l) has been set to 0.2, the following formula is used:

$$w_{GK} = w_{GK} + l \times Error_K \times Output_G$$
$$w_{GK} = 0.3 + 0.2 \times -0.064 \times 0.84$$
$$w_{GK} = 0.276$$

In this example, the weight has been adjusted lower. The remaining weights in the network are adjusted and the process continues with another example presented to the neural network causing the weights of all connections in the network to be adjusted based on the calculated error values.

The following example works through the entire process of how a neural network learns from a single training example. Figure 7.33 shows the first eight steps of the learning process. A normalized training set of observations will be used in the learning process. This training set has three input descriptor variables ($\mathbf{I}_1$, $\mathbf{I}_2$, and $\mathbf{I}_3$) and one output response variable ($\mathbf{O}$). Five observations are shown (i, ii, iii, iv, and v). In step 1, a neural network is set up, which has three input nodes (A, B, and C) corresponding to each of the three input variables and one output node (F), since there is only one output response variable. The neural network has a single hidden layer consisting of two nodes (D and E). All nodes from the input layer are connected to the two hidden layer nodes. These two nodes are in turn connected to the output layer, which is the single node F in this example. In addition to setting up the structure or *topology* of the network, random numbers between -1 and $+1$ are assigned as weights to each of the connections. For example, the weight from node A to node D (w_{AD}) is 0.4 and the weight from node E to F (w_{EF}) is 0.1.

In step 2, a random observation is selected (v) and is presented to the network as shown. The value of $\mathbf{I}_1$ (0) is presented to A, the value of $\mathbf{I}_2$ (1) is presented to B, and the value of $\mathbf{I}_3$ (1) is presented to C. In step 3, these inputs in combination with the connection weights are used to calculate the output from the hidden nodes D and E. To calculate these outputs, nodes D and E first combine the inputs and then use an activation function to derive the outputs. The combined inputs to nodes D and E are the weighted sum of the input values:

$$Input_D = I_1 \times w_{AD} + I_2 \times w_{BD} + I_3 \times w_{CD}$$
$$Input_D = (0 \times 0.4) + (1 \times -0.6) + (1 \times 0.9) = 0.3$$

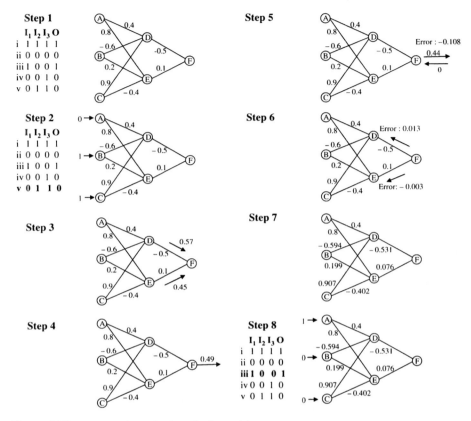

Figure 7.33. Process of learning by adjusting weights

$$Input_E = I_1 \times w_{AE} + I_2 \times w_{BE} + I_3 \times w_{CE}$$
$$Input_E = (0 \times 0.8) + (1 \times 0.2) + (1 \times -0.4) = -0.2$$

The outputs from D and E use these combined input values within an activation function to generate the output values. In this case, we will use a sigmoid activation function:

$$Output_D = \frac{1}{1 + e^{-Input_D}} = \frac{1}{1 + e^{-0.3}} = 0.57$$
$$Output_E = \frac{1}{1 + e^{-Input_E}} = \frac{1}{1 + e^{0.2}} = 0.45$$

In step 4, the outputs from D and E are used as inputs to F. The total input is calculated by combining these values with the weights of the connections:

$$Input_F = Output_D \times w_{DF} + Output_E \times w_{EF}$$
$$Input_F = (0.57 \times -0.5) + (0.45 \times 0.1) = -0.24$$

Next, this input is converted to an output value using the activation function:

$$Output_F = \frac{1}{1 + e^{-Input_F}} = \frac{1}{1 + e^{0.24}} = 0.44$$

In step 5, the value calculated as the output from node F is now compared to the actual output value. An error value is computed:

$$Error_F = Output_F(1 - Output_F)(Actual_F - Output_F)$$
$$Error_F = 0.44(1 - 0.44)(0 - 0.44) = -0.108$$

In step 6, using the error calculated for node F, an error is calculated for nodes D and E:

$$Error_D = Output_D(1 - Output_D)(Error_F \times w_{DF})$$
$$Error_D = 0.57(1 - 0.57)(-0.108 \times -0.5) = 0.013$$
$$Error_E = Output_E(1 - Output_E)(Error_F \times w_{EF})$$
$$Error_E = 0.45(1 - 0.45)(-0.108 \times 0.1) = -0.003$$

In step 7, these error calculations for nodes D, E, and F can now be used to calculate the new weights for the network. A constant learning rate (l) of 0.5 will be used in the following equations:

$$w_{ij} = w_{ij} + l \times Error_j \times Output_i$$
$$w_{AD} = 0.4 + 0.5 \times 0.013 \times 0 = 0.4$$
$$w_{AE} = 0.8 + 0.5 \times -0.003 \times 0 = 0.8$$
$$w_{BD} = -0.6 + 0.5 \times 0.013 \times 1 = -0.594$$
$$w_{BE} = 0.2 + 0.5 \times -0.003 \times 1 = 0.199$$
$$w_{CD} = 0.9 + 0.5 \times 0.013 \times 1 = 0.907$$
$$w_{CE} = -0.4 + 0.5 \times -0.003 \times 1 = -0.402$$
$$w_{DF} = -0.5 + 0.5 \times -0.108 \times 0.57 = -0.531$$
$$w_{EF} = 0.1 + 0.5 \times -0.108 \times 0.45 = 0.076$$

The weights in the network have been adjusted and a new random example is presented to the network in step 8 (observation iii) and the process of learning continues.

7.5.7 Using Neural Networks

When learning from a training set, there are a number of parameters to adjust that influence the quality of the prediction, including the following:

- **Hidden layers:** Both the number of hidden layers and the number of nodes in each hidden layer can influence the quality of the results. For example, too few layers and/or nodes may not be adequate to sufficiently learn and too many may result in overtraining the network.

- **Number of cycles:** A cycle is where a training example is presented and the weights are adjusted. The number of examples that get presented to the neural network during the learning process can be set. The number of cycles should be set to ensure that the neural network does not overtrain. The number of cycles is often referred to as the number of *epochs*.

- **Learning rate:** Prior to building a neural network, the learning rate should be set and this influences how fast the neural network learns.

Neural networks have a number of advantages:

- **Linear and nonlinear models:** Complex linear and nonlinear relationships can be derived using neural networks.

- **Flexible input/output:** Neural networks can operate using one or more descriptors and/or response variables. They can also be used with categorical and continuous data.

- **Noise:** Neural networks are less sensitive to noise than statistical regression models.

The major drawbacks with neural networks are:

- **Black box:** It is not possible to explain how the results were calculated in any meaningful way.

- **Optimizing parameters:** There are many parameters to be set in a neural network and optimizing the network can be challenging, especially to avoid overtraining.

7.5.8 Example

When building a neural network, it is important to optimize the network to generate a good prediction at the same time as ensuring the network is not overtrained. The following example illustrates the use of neural networks in prediction using the automobile example. A 10% cross validation method was used to assess the models built.

Neural networks were built using the following parameters:

Inputs: **Horsepower**, **Weight**, **Model Year**, and **Origin**

Output: **MPG**

Hidden layers: 2

Learning rate: 0.2

Figure 7.34 illustrates the learning process. The neural network was run for 5,000, 10,000, and 20,000 cycles. The scatterplot shows the relationship between the actual values and the predictions along with the r^2 values for these relationships. It can be seen from the three scatterplots that as the number of cycles increases, the accuracy of the model increases. This can also be seen in Figure 7.35. The chart has plotted a series of models generated using different numbers of cycles. The x-axis shows the

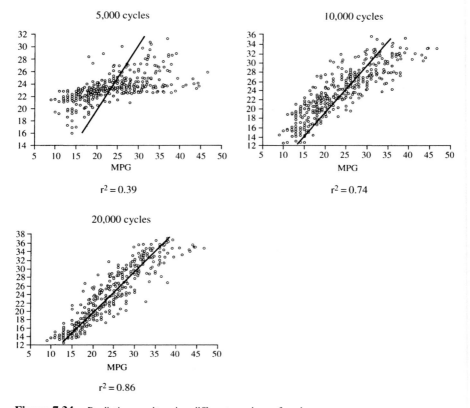

Figure 7.34. Prediction results using different numbers of cycles

number of cycles used for each model and the *y*-axis shows the cross-validated r^2 value for each model. It illustrates how the neural network rapidly learns initially. As the network approaches the optimal accuracy, it learns more slowly. Eventually the network will start to overlearn and will not be able to generalize as well for examples outside the training set. This can be tested using a separate test set. If the predictive accuracy of the neural network starts to decline, the network is overtraining.

To identify an optimal neural network for prediction, an experiment is designed to test three parameters:

- **The inputs:** All combinations of descriptors were tested from two to seven. The descriptors used were **Cylinders** (C), **Displacement** (D), **Weight** (W), **Acceleration** (A), **Model Year** (MY), and **Origin** (O). For example, where the inputs were **Weight, Model Year**, and **Origin** these were designated as W, MY, O.

- **The number of hidden layers:** One and two hidden layers were tested to observe the impact on the predictive accuracy of the model. A more extensive experiment may test additional neural network topologies.

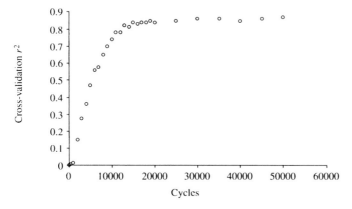

Figure 7.35.   Neural network learning as cycles increase

- **The number of cycles:** For each combination of inputs and number of hidden layers, a series of models were built using 1,000, 5,000, 10,000, and 25,000 cycles. These values were selected in order to understand the curve as shown in Figure 7.35. As the number of cycles increases, the predictive accuracy of the model increases towards an optimal accuracy.

Table 7.19 shows the r^2 values for different combinations of the three parameters. The objective when selecting the parameters is to keep the model as simple as possible with the fewest number of inputs and the smallest number of hidden layers at the same time as ensuring the model has not overtrained. The following model parameters were chosen to build the final model. The model built using these parameters has one of the highest r^2 values:

Inputs: **Horsepower, Weight, Model Year,** and **Origin**

Cycles: 10,000

Hidden layers: 1

Learning rate: 0.2

The multilayer backpropagation neural network, as presented here, is one type of network. Other approaches are referenced in the further reading section of this chapter.

7.6   OTHER METHODS

There are many methods for building both classification and regression models. The following section briefly describes a number of alternative approaches. More details on these approaches are referenced in the further reading section of this chapter.

- **Multiple linear regressions:** The method described for simple linear regression can be extended to handle multiple descriptor variables. A similar least squares method is used to generate the equation. The form of the equation is

Table 7.19. Optimization of neural network

Inputs	Layers	1k	5k	10k	25k	Inputs	Layers	1k	5k	10k	25k	Inputs	Layers	1k	5k	10k	25k
			cycles						cycles						cycles		
C, D, H, W, A, MY, O	1	0.65	0.809	0.827	0.855	C, A, MY, O	1	0.42	0.725	0.735	0.742	D, MY, O	1	0.256	0.588	0.714	0.788
	2	0.316	0.768	0.822	0.853		2	0.009	0.441	0.729	0.738		2	0.01	0.301	0.539	0.729
D, H, W, A, MY, O	1	0.627	0.815	0.847	0.857	C, W, MY, O	1	0.48	0.772	0.817	0.841	H, W, A	1	0.151	0.475	0.661	0.738
	2	0.096	0.686	0.839	0.855		2	0.112	0.413	0.724	0.83		2	0.0	0.021	0.264	0.566
C, H, W, A, MY, O	1	0.67	0.802	0.834	0.853	C, W, A, O	1	0.202	0.624	0.693	0.715	H, W, O	1	0.221	0.581	0.684	0.733
	2	0.212	0.719	0.819	0.853		2	0.017	0.38	0.514	0.691		2	0.001	0.074	0.379	0.667
C, D, W, A, MY, O	1	0.547	0.793	0.811	0.843	C, W, A, MY	1	0.541	0.803	0.812	0.848	H, A, MY	1	0.034	0.519	0.704	0.743
	2	0.084	0.709	0.792	0.837		2	0.134	0.378	0.77	0.834		2	0.0	0.128	0.282	0.628
C, D, H, A, MY, O	1	0.598	0.792	0.79	0.826	C, H, MY, O	1	0.53	0.759	0.797	0.81	H, MY, O	1	0.578	0.602	0.73	0.768
	2	0.146	0.642	0.784	0.813		2	0.009	0.342	0.691	0.797		2	0.004	0.181	0.472	0.725
C, D, H, W, MY, O	1	0.615	0.819	0.844	0.86	C, H, A, O	1	0.434	0.629	0.662	0.708	W, A, MY	1	0.24	0.737	0.788	0.846
	2	0.072	0.768	0.813	0.852		2	0.002	0.357	0.541	0.7		2	0.002	0.245	0.392	0.76
C, D, H, W, A, O	1	0.506	0.693	0.709	0.725	C, H, A, MY	1	0.482	0.748	0.763	0.787	W, MY, O	1	0.35	0.715	0.799	0.839
	2	0.114	0.615	0.693	0.722		2	0.027	0.302	0.539	0.785		2	0.004	0.171	0.594	0.821
C, D, H, W, A, MY	1	0.585	0.811	0.838	0.853	C, H, W, O	1	0.485	0.651	0.69	0.75	A, MY, O	1	0.145	0.514	0.531	0.545
	2	0.096	0.764	0.826	0.852		2	0.047	0.307	0.644	0.732		2	0.011	0.235	0.356	0.508
H, W, A, MY, O	1	0.625	0.792	0.848	0.864	C, H, W, MY	1	0.597	0.782	0.828	0.853	C, D	1	0.136	0.557	0.583	0.545
	2	0.098	0.63	0.783	0.865		2	0.006	0.778	0.774	0.848		2	0.021	0.048	0.303	0.268
D, W, A, MY, O	1	0.578	0.801	0.831	0.845	C, H, W, A	1	0.458	0.671	0.698	0.731	C, H	1	0.016	0.429	0.549	0.589
	2	0.097	0.58	0.811	0.848		2	0.072	0.339	0.658	0.708		2	0.008	0.016	0.259	0.383
D, H, A, MY, O	1	0.393	0.78	0.812	0.826	C, D, MY, O	1	0.553	0.754	0.765	0.772	C, W	1	0.233	0.516	0.593	0.596
	2	0.011	0.55	0.765	0.833		2	0.13	0.544	0.752	0.765		2	0.021	0.004	0.158	0.423

Subset					
D, H, W, MY, O	1	0.536	0.825	0.844	0.863
	2	0.081	0.684	0.835	0.851
D, H, W, A, O	1	0.561	0.688	0.712	0.741
	2	0.096	0.601	0.683	0.736
D, H, W, A, MY	1	0.594	0.833	0.846	0.857
	2	0.077	0.541	0.733	0.856
C, W, A, MY, O	1	0.578	0.792	0.811	0.844
	2	0.126	0.666	0.792	0.842
C, H, A, MY, O	1	0.539	0.778	0.788	0.802
	2	0.066	0.664	0.742	0.799
C, H, W, MY, O	1	0.614	0.801	0.829	0.86
	2	0.018	0.581	0.792	0.853
C, H, W, A, O	1	0.524	0.674	0.702	0.736
	2	0.07	0.548	0.658	0.728
C, H, W, A, MY	1	0.66	0.803	0.833	0.849
	2	0.107	0.659	0.802	0.85
C, D, A, MY, O	1	0.542	0.741	0.767	0.773
	2	0.114	0.683	0.739	0.769
C, D, W, MY, O	1	0.585	0.802	0.81	0.844
	2	0.069	0.666	0.79	0.836
C, D, W, A, O	1	0.567	0.655	0.69	0.709
	2	0.077	0.639	0.644	0.706
C, D, W, A, MY	1	0.475	0.795	0.825	0.849
	2	0.108	0.592	0.808	0.841
C, D, H, MY, O	1	0.484	0.778	0.801	0.821
	2	0.175	0.647	0.787	0.811

Subset					
C, D, A, O	1	0.182	0.621	0.655	0.648
	2	0.09	0.387	0.546	0.631
C, D, A, MY	1	0.536	0.715	0.759	0.766
	2	0.004	0.344	0.726	0.778
C, D, W, O	1	0.325	0.653	0.674	0.707
	2	0.029	0.286	0.549	0.695
C, D, W, MY	1	0.471	0.801	0.823	0.844
	2	0.008	0.489	0.718	0.824
C, D, W, A	1	0.516	0.682	0.696	0.718
	2	0.033	0.232	0.572	0.71
C, D, H, O	1	0.419	0.612	0.689	0.732
	2	0.032	0.397	0.622	0.72
C, D, H, A	1	0.407	0.665	0.683	0.721
	2	0.005	0.32	0.553	0.718
C, D, H, W	1	0.581	0.702	0.718	0.744
	2	0.009	0.395	0.515	0.704
C, D, H	1	0.33	0.624	0.621	0.717
	2	0	0.078	0.456	0.6
C, D, W	1	0.237	0.624	0.634	0.712
	2	0.047	0.259	0.477	0.628
C, D, A	1	0.259	0.464	0.642	0.659
	2	0.01	0.115	0.395	0.629
C, D, MY	1	0.256	0.737	0.751	0.783
	2	0.005	0.282	0.491	0.7
C, D, O	1	0.461	0.557	0.629	0.663
	2	0.002	0.109	0.504	0.611

Subset					
C, A	1	0.095	0.524	0.531	0.532
	2	0.0	0.017	0.096	0.274
C, MY	1	0.013	0.629	0.658	0.656
	2	0.004	0.117	0.2	0.446
C, O	1	0.188	0.492	0.498	0.481
	2	0.0	0.185	0.225	0.296
D, H	1	0.033	0.554	0.589	0.643
	2	0.0	0.054	0.048	0.295
D, W	1	0.342	0.605	0.622	0.641
	2	0.0	0.092	0.251	0.247
D, A	1	0.044	0.549	0.559	0.564
	2	0.01	0.048	0.167	0.322
D, MY	1	0.144	0.594	0.647	0.679
	2	0.001	0.041	0.31	0.295
D, O	1	0.032	0.482	0.493	0.49
	2	0.007	0.111	0.206	0.324
H, W	1	0.015	0.649	0.682	0.642
	2	0.0	0.019	0.186	0.232
H, A	1	0.004	0.381	0.439	0.462
	2	0.002	0.006	0.098	0.133
H, MY	1	0.059	0.517	0.548	0.587
	2	0.004	0.007	0.078	0.341
H, O	1	0.037	0.422	0.475	0.478
	2	0.003	0.045	0.155	0.353
W, A	1	0.144	0.553	0.597	0.59
	2	0.0	0.0	0.084	0.313

Table 7.19. (*Continued*)

Inputs	Layers	cycles 1k	5k	10k	25k
C, D, H, A, O	1	0.588	0.667	0.685	0.704
	2	0.086	0.522	0.64	0.679
C, D, H, A, MY	1	0.43	0.649	0.682	0.716
	2	0.124	0.524	0.652	0.71
C, D, H, W, O	1	0.571	0.671	0.721	0.747
	2	0.192	0.596	0.678	0.723
C, D, H, W, MY	1	0.6	0.812	0.83	0.853
	2	0.204	0.77	0.826	0.848
C, D, H, W, A.	1	0.543	0.686	0.716	0.728
	2	0.161	0.544	0.582	0.724
W, A, MY, O	1	0.254	0.776	0.827	0.858
	2	0.081	0.467	0.765	0.848
H, A, MY, O	1	0.377	0.703	0.766	0.814
	2	0.027	0.266	0.378	0.813
H, W, MY, O	1	0.412	0.786	0.85	0.865
	2	0.042	0.55	0.675	0.859
H, W, A, O	1	0.359	0.669	0.708	0.728
	2	0.052	0.305	0.641	0.733
H, W, A, MY	1	0.54	0.806	0.839	0.864
	2	0.024	0.382	0.817	0.861
D, A, MY, O	1	0.448	0.715	0.754	0.789
	2	0.008	0.541	0.466	0.796

Inputs	Layers	cycles 1k	5k	10k	25k
C, H, W	1	0.468	0.595	0.703	0.744
	2	0.019	0.262	0.413	0.669
C, H, A	1	0.222	0.521	0.632	0.693
	2	0.01	0.246	0.302	0.627
C, H, MY	1	0.423	0.59	0.758	0.788
	2	0.0	0.163	0.583	0.739
C, H, O	1	0.321	0.573	0.646	0.719
	2	0.001	0.204	0.359	0.532
C, W, A	1	0.182	0.51	0.676	0.695
	2	0.025	0.226	0.453	0.656
C, W, MY	1	0.386	0.784	0.814	0.823
	2	0.005	0.273	0.647	0.811
C, W, O	1	0.305	0.538	0.664	0.691
	2	0.005	0.166	0.411	0.649
C, A, MY	1	0.252	0.656	0.701	0.711
	2	0.032	0.199	0.5	0.698
C, A, O	1	0.401	0.531	0.584	0.623
	2	0.002	0.166	0.257	0.612
C, MY, O	1	0.354	0.69	0.735	0.742
	2	0.021	0.217	0.342	0.67
D, H, W	1	0.402	0.549	0.7	0.744
	2	0.064	0.096	0.464	0.652

Inputs	Layers	cycles 1k	5k	10k	25k
W, MY	1	0.021	0.616	0.665	0.693
	2	0.004	0.019	0.042	0.485
W, O	1	0.054	0.489	0.493	0.469
	2	0.005	0.044	0.135	0.288
A, MY	1	0.001	0.334	0.336	0.343
	2	0.02	0.028	0.07	0.237
A, O	1	0.001	0.309	0.386	0.394
	2	0.009	0.015	0.039	0.249
MY, O	1	0.177	0.522	0.53	0.514
	2	0.002	0.021	0.123	0.331

D. W. MY. O	1	0.424	0.651	0.696	0.712
	2	0.028	0.308	0.581	0.696
D. W. A. O	1	0.42	0.627	0.703	0.718
	2	0.004	0.32	0.421	0.708
D. W. A. MY	1	0.464	0.821	0.832	0.848
	2	0.061	0.275	0.782	0.844
D. H. MY. O	1	0.484	0.76	0.797	0.833
	2	0.048	0.379	0.564	0.83
D. H. A. O	1	0.362	0.636	0.683	0.726
	2	0.084	0.345	0.611	0.719
D. H. W. MY	1	0.528	0.832	0.855	0.86
	2	0.027	0.399	0.789	0.847
D. H. W. A	1	0.325	0.711	0.728	0.743
	2	0.015	0.391	0.588	0.714

D. H. A	1	0.353	0.593	0.608	0.726
	2	0.0	0.278	0.369	0.666
D. H, MY	1	0.311	0.714	0.799	0.82
	2	0.002	0.028	0.345	0.739
D. H, O	1	0.249	0.567	0.669	0.728
	2	0.004	0.186	0.348	0.652
D, W, A	1	0.087	0.568	0.708	0.72
	2	0.009	0.175	0.264	0.631
D, W, MY	1	0.108	0.814	0.841	0.843
	2	0.015	0.288	0.498	0.793
D. W, O	1	0.382	0.555	0.671	0.714
	2	0.011	0.207	0.416	0.7
D. A. MY	1	0.175	0.604	0.778	0.79
	2	0.005	0.066	0.341	0.76

$y = a + b_1 x_1 + b_2 x_2 + \cdots + b_n x_k$ where y is the response, x_1 to x_n are the descriptor variables, a is a constant, and b_1 to b_n are also constants. For example, when attempting to predict a potential customer's credit score (**CS**) a multiple linear regression equation could be generated. The equation could be based on the number of missed payments to other credit cards (**MP**), the number of years with no missed payments (**NMP**), and the number of good standing loans (**GSL**), as for example in the following equation:

$$\textbf{CS} = 15 - 18 \times \textbf{MP} + 12 \times \textbf{NMP} + 10 \times \textbf{GSL}$$

- **Logistic regression:** Logistic regression is a regression method that can be applied in situations where the response variable is dichotomous and usually translated to a binary variable.
- **Random forests:** A random forest is a collection of decision trees used to make predictions. Each tree is built using a subset of all observations. In addition, each tree is built with a subset of all possible descriptors to use in splitting the observations. When using a random forest to make predictions, the observation is presented to each tree. Each individual tree makes a prediction and the results from all the trees are combined to create a final prediction using either an average or a voting scheme in the case of categorical responses.
- **Rule-based classifiers:** In Chapter 6, a number of methods were described that generate rules from the data, for example, associative rules. When the THEN-part of the rule is a response variable, these rules can be used to build classification models. When a new observation is presented to the model, rules are identified that match the IF-part of the rule to the observation. The predicted classification corresponds to the THEN-part of the rule. If multiple rules match a single observation, then either the rule with the highest confidence is selected or a voting scheme is used. Rule-based classifiers provide a quick method of classification that is easy to interpret.
- **Naïve Bayes classifiers:** This is a method of classification that makes use of Bayes theorem. It assumes that the descriptor variables used are independent. The method is capable of handling noise and missing values.
- **Partial least squares regression:** Partial least squares regression combines multiple linear regressions and principal component analysis. It can be used to handle nonlinear multiple-regression problems.
- **Support vector machines:** Support vector machines can be used for both classification and regression problems. For classification problems, they attempt to identify a hyperplane that separates the classes. Despite their general usefulness, they can be difficult to interpret.

7.7 SUMMARY

Types of models:

- **Classification:** Models where the response variable is categorical. These are assessed using: concordance, error rate, specificity, and sensitivity analysis.

Table 7.20. Summary of predictive modeling approaches in this chapter

	Response type	Descriptor type	Problem	Time to build	Time to apply	Explanation
Simple linear regression	Single continuous	Single continuous	Linear	Fast	Fast	Formula
kNN	Single any	Any	Based on similar observations	Fast	Slow	Similar observations
Regression trees	Single continuous	Any	Based on property ranges	Slow	Fast	Tree
Classification trees	Single categorical	Any	Based on property ranges	Slow	Fast	Tree
Neural nets	Any	Any	Nonlinear	Slow	Fast	No explanation

- **Regression:** Models where the response variable is continuous. These are assessed using r^2 and residual analysis.

Building a prediction model involves the following steps:

1. Select methods based on problem
2. Separate out training and test sets
3. Optimize the models
4. Assess models generated

Applying a prediction model follows these steps:

1. Evaluate whether an observation can be used with the model
2. Present observations to model
3. Combine results from multiple models (if appropriate)
4. Understand confidence and/or explain how results were computed

Table 7.20 summarizes the different methods described in this chapter.

7.8 EXERCISES

1. A classification prediction model was built using a training set of examples. A separate test set of 20 examples is used to test the model. Table 7.21 shows the results of applying this test set. Calculate the model's:
   a. Concordance
   b. Error rate
   c. Sensitivity
   d. Specificity

Table 7.21. Table of actual vs predicted values
(categorical response)

Observation	Actual	Predicted
1	0	0
2	1	1
3	1	1
4	0	0
5	0	0
6	1	0
7	0	0
8	0	0
9	1	1
10	1	1
11	1	1
12	0	1
13	0	0
14	1	1
15	0	0
16	1	1
17	0	0
18	1	1
19	0	1
20	0	0

2. A regression prediction model was built using a training set of examples. A separate test set was applied to the model and the results are shown in Table 7.22.

   a. Determine the quality of the model using r^2
   b. Calculate the residual for each observation

3. Table 7.23 shows the relationship between the amount of fertilizer used and the height of a plant.

   a. Calculate a simple linear regression equation using **Fertilizer** as the descriptor and **Height** as the response.
   b. Predict the height when fertilizer is 12.3

4. A kNN model is being used to predict house prices. A training set was used to generate a kNN model and k is determined to be 5. The unseen observation in Table 7.24 is presented to the model. The kNN model determines the five observations in Table 7.25 from the training set to be the most similar. What would be the predicted house price value?

5. A classification tree model is being used to predict which brand of printer a customer would purchase with a computer. The tree in Figure 7.36 was built from a training set of examples. For a customer whose **Age** is 32 and **Income** is $35,000, which brand of printer would the tree predict he/she would buy?

6. Figure 7.37 shows a simple neural network. An observation with two variables (0.8, 0.2) is presented to the network as shown. What is the predicted output from the neural network using a sigmoid activation function?

Table 7.22. Table of actual vs predicted values
(continuous response)

Observation	Actual	Predicted
1	13.7	12.4
2	17.5	16.1
3	8.4	6.7
4	16.2	15.7
5	5.6	8.4
6	20.4	15.6
7	12.7	13.5
8	5.9	6.4
9	18.5	15.4
10	17.2	14.5
11	5.9	5.1
12	9.4	10.2
13	14.8	12.5
14	5.8	5.4
15	12.5	13.6
16	10.4	11.8
17	8.9	7.2
18	12.5	11.2
19	18.5	17.4
20	11.7	12.5

Table 7.23. Table of plant experiment

Fertilizer	Height
10	0.7
5	0.4
12	0.8
18	1.4
14	1.1
7	0.6
15	1.3
13	1.1
6	0.6
8	0.7
9	0.7
11	0.9
16	1.3
20	1.5
17	1.3

Table 7.24.   House with unknown price

Bedroom	Number of bathrooms	Square feet	Garage	House price
2	2	1,810	0	

Table 7.25.   Table of similar observations

Bedroom	Number of bathrooms	Square feet	Garage	House price
2	2	1,504	0	355,000
2	2	1,690	0	352,000
2	3	1,945	0	349,000
3	2	2,146	0	356,000
3	2	1,942	0	351,000

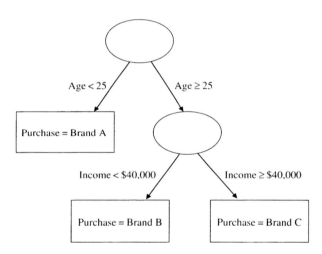

Figure 7.36.   Classification tree for customer's brand purchase based on **Age** and **Income**

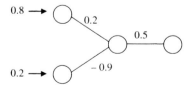

Figure 7.37.   Simple neural network

7.9 FURTHER READING

For more information on methods for combining models, such as bagging and boosting, see Witten (2000), and for confidence metrics for simple linear regression see Kachigan (1991), Donnelly (2004), and Levine (2005). Fausett (1994) provides additional details on Neural Networks. The following indicates sources of additional information on the following topics: multiple linear regression (Kleinbaum 1998), logistic regression (Agresti 2002), random forests (Kwok 1990), rule-based classifiers (Tang 2005), Naïve Bayes (Tang 2005), partial least squares regressions (Wold 1975), and support vector machines (Cristianini 2000).

Chapter 8

Deployment

8.1 OVERVIEW

To realize the benefits of a data analysis or data mining project, the solution must be deployed, that is, applied to the business or scientific problem. It is important to plan this part of the project to ensure that the analysis performed to date positively influences the business. The following sections briefly outline deliverables and activities necessary during any deployment step.

8.2 DELIVERABLES

There are many options for delivering data analysis or data mining solutions. Some of the more popular include:

- **Report:** A report describing the business intelligence derived from the project is a common deliverable. The report should be directed to the persons responsible for making decision. It should focus on significant and actionable items, that is, it should be possible to translate any conclusions into a decision that can be used and that makes a difference. It is increasingly common for the report to be delivered through the corporate intranet to enable additional interested parties to benefit from the report.

- **Integration into existing systems:** The integration of the results into existing operational systems or databases is often one of the most cost effective approaches to delivering a solution. For example, when a sales team requires the results of a predictive model, that ranks potential customers on the basis of the likeliness that they will buy a particular product, the model may be integrated with the CRM system (Customer Relationship Management) that they currently use on a daily basis. This minimizes the need for training and makes the deployment of the results easier. Prediction models or data mining results can also be integrated into system accessible to a customers such as e-commerce web sites. In this situation, customers may be presented with additional products or services that they may be interested

Making Sense of Data: A Practical Guide to Exploratory Data Analysis and Data Mining,
By Glenn J. Myatt
Copyright © 2007 John Wiley & Sons, Inc.

in, identified using an embedded prediction model. Models may need to be integrated into existing operational processes where a model needs to be constantly applied to operational data. For example, a solution may be the detection of events leading to errors in a manufacturing system. Catching these issues early enough may allow a technician to rectify the problem without stopping the production system. The model needs to be integrated with the data generated from the system. Any identified anomalies should be rapidly communicated to those who might be able to prevent the potential problem. A data mining solution may also require continual access to new training data since the data from which a model is built is only relevant for a short period of time. In this situation, it will be essential to tightly integrate the model building with the data. Core technologies involved in the deployment include tools used to perform the analysis (statistics, OLAP, visualizations and data mining), methods for sharing the models generated, integration with databases and workflow management systems. The further reading section of this chapter provides links to resources on deploying data analysis and data mining solutions.

- **Standalone software:** Another option is the development of a standalone system. The advantage of this approach is that, since it is not necessary to integrate with operational systems, the solution may be deployed more rapidly. However, there is a cost in terms of developing, maintaining and training.

8.3 ACTIVITIES

The following activities need to be accomplished during the deployment phase:

- **Plan and execute the deployment:** A plan should be generated describing the deployment of the solutions. It should include information on how the solution is to be deployed and to whom. Issues relating to the management of change, as the solution may introduce changes to some individual's daily activities, should be addressed. Also, a deployment may require a training program that outlines both how to use the new technology and how to interpret the results. In many situations the value of the data, and hence the models generated from the data, diminishes over time. In this situation, updated models may be required and a strategy should be put in place to ensure the currency of the models. This could be accomplished through an automated approach or through manually updating of the models and needs to be planned.

- **Measure and monitor performance:** It is important to understand if the models or analysis generated translate into meeting the business objectives outlined at the start of the project. For example, the models may be functioning as expected; however, the individuals that were expected to use the solution are not for some reasons and hence there is no business benefit. A controlled experiment (ideally double blind) in the field should be considered to assess the quality of the results and their business impact. For example, the intended users of a predictive model could be divided into two groups. One group, made up of half (randomly selected) of the users, uses the model results and the other group does not. The business impact resulting from the two groups could then be measured. When models are continually updated, the consistency of the results generated should be also monitored over time.

- **Review project:** At the end of a project, it is always a useful exercise to look back at what worked and what did not work. This will provide insights to improve future projects.

8.4 DEPLOYMENT SCENARIOS

Exploratory data analysis and data mining has been deployed to a variety of problems. The following illustrates some of the areas where this technology has been deployed:

- **Personalized e-commerce:** Customers characteristics, based on profiles and historical purchasing information, can be used to personalize e-commerce web sites. Customers can be directed to products and services matching their anticipated needs.

- **Churn analysis:** Profiles of customers discontinuing a particular product or service can be analyzed and prediction models generated for customers who are likely to switch. These models can be used to identify at risk customers providing an opportunity to target them with a focused marketing campaign in order to retain their business.

- **Quality control:** Quality is critical to all production systems and exploratory data analysis and data mining approaches are important tools in creating and maintaining a high quality production system. For example, the 6-sigma quality control methodology uses many of the statistical methods described in Chapter 5.

- **Experimental design and analysis:** Experiments are widely used in all areas of research and development to design, test and assess new products. Exploratory data analysis and data mining are key tools in both the design of these experiments and the analysis of the results. For example, every day biologists are experimentally generating millions of data points concerning genes and it is critical to make use of exploratory data analysis and data mining in order to make sense out of this data.

- **Targeted marketing campaigns:** Organizations can use data analysis and data mining methods to understand profiles of customers who are more likely to purchase specific products and use this information for more targeted marketing campaigns with higher response rates.

- **Analyzing the results of surveys:** Surveys are a widely used way of determining opinions and trends in the market place, and the application of exploratory data analysis and data mining to the process of collecting and analyzing the result will help to get the answers faster.

- **Anomaly detection:** In many situations it is the detection of outliers in the data that is most interesting. For example, the detection of fraudulent insurance claim applications can be based on the analysis of unusual activity.

8.5   SUMMARY

Table 8.1 summarizes issues to consider when deploying any solution.

8.6   FURTHER READING

The following web sites provide a list of some of the tools for deploying data analysis and/or data mining solutions:

http://www.angoss.com/

http://www.fairisaac.com/

http://www-306.ibm.com/software/data/iminer/

http://www.insightful.com/

http://www.jmp.com/

http://www.kxen.com/

http://www.microsoft.com/sql/solutions/bi/default.mspx

Table 8.1.   Deployment issues

Deliverables	Report	Describes significant and actionable items
	Integration into existing systems	Cost effective solution, minimal training cost, minimal deployment cost, access to up-to-date information
	Standalone software	May provide rapid deployment
Activities	Plan and execute deployment	Describes how and to whom the solution will be deployed, identify whether there is a need for change management, describes any required training, discusses how the models will be kept up-to-date
	Measure and monitor performance	Determines to what degree the project has met the success criteria, ensures that the model results are consistent over time
	Review project	To understand what worked and what did not work

http://www.oracle.com/technology/products/bi/odm/index.html

http://www.sas.com/index.html

http://www.spss.com/

http://www.statsoft.com/

http://www.systat.com/

The following resources provide support for integrating data mining solutions:

http://www.dmg.org/

http://www.jcp.org/en/jsr/detail?id=73

The following references provide additional case studies: Guidici (2005), Rudd (2001) and Berry (2004).

Chapter 9

Conclusions

9.1 SUMMARY OF PROCESS

Exploratory data analysis and data mining is a process involving defining the problem, collecting and preparing the data, and implementing the analysis. Once completed and evaluated, the project should be delivered to the consumer concerned by the information. Following a process has many advantages including avoiding common pitfalls in analyzing data and ensuring that the project meets expectations. This book has described the process in four steps:

1. **Problem definition:** Prior to any analysis, the problem to be solved should be clearly defined and related to one or more business objectives. Describing the deliverables will focus the team on delivering the solution and provides correct expectations to other parties interested in the outcome of the project. A multidisciplinary team is best suited to solve these problems driven by a project leader. A plan for the project should be developed, covering the objectives and deliverables along with a timeline and a budget. An analysis of the relationship between the cost of the project and the benefit derived for the business can form a basis for a go/no-go decision for the project.

2. **Data preparation:** The quality of the data is the most important aspect that influences the quality of the results from the analysis. The data should be carefully collected, integrated, characterized, and prepared for analysis. Data preparation includes cleaning the variables to ensure consistent naming and removing potential errors. Eliminating variables that provide little benefit to any analysis can be done at this stage. The variables should be characterized and potentially transformed to ensure that the variables are considered with equal weight, that they match as closely as possible a normal distribution, and also to enable the use of the data with multiple analysis methods. Where appropriate, the data set should be partitioned into smaller sets to simplify the analysis. At the

Making Sense of Data: A Practical Guide to Exploratory Data Analysis and Data Mining,
By Glenn J. Myatt
Copyright © 2007 John Wiley & Sons, Inc.

end of the data preparation phase, one should be very familiar with the data and should already start identifying aspects of the data relating to the problem being solved. The steps performed in preparing the data should be documented. A data set ready for analysis should have been prepared.

3. **Implementation of the analysis:** There are three primary tasks that relate to any data analysis or data mining project: summarizing the data, finding hidden relationships, and making predictions. When implementing the analysis one should select appropriate methods that match the task, the data, and the objectives of the project. Available methods include graphing the data, summarizing the data in tables, descriptive statistics, inferential statistics, correlation analysis, grouping methods, and mathematical models. Graphs, summary tables, and descriptive statistics are essential for summarizing data. Where general statements about populations are needed, inferential statistics should be used to understand the statistical significance of the summaries. Where a method is being used for grouping or prediction, appropriate methods should be selected that match the objectives of the projects and the available data. These methods should be fined-tuned, adjusting the parameters within a controlled experiment. When assessing the quality of a prediction model, a separate test and training set should be used. When presenting the results of the analysis, any transformed data should be presented in its original form. Appropriate methods for explaining and qualifying the results should be developed when needed. Where an analysis is based on multiple models, specific model selection criteria and/or composite models should be developed.

4. **Deployment:** A plan should be set up to deliver the results of the analysis to the already identified consumer. This plan will need to take into account nontechnical issues of introducing a solution that potentially changes the user's daily routine. The plan may need to address the need for continual updates to the predictive models over time. The plan should be executed as well as the performance measured. This performance should directly relate to the business objectives of the project. This performance may change over time and should be monitored.

Although the process is described as a linear four-step approach, most projects will invariably need to go between the different stages from time-to-time. Like any complex technical project, this process needs to be managed by a project leader to ensure that the project is planned and delivered on time. Communication between the cross-disciplinary teams and other stakeholders about progress is essential. Regular status meeting, especially between steps in the process, are critical. Table 9.1 summarizes the process.

Three issues for delivering a successful project should be highlighted. Firstly, a clear and measurable objective will help to focus the project on issues that

Table 9.1. Table summarizing process and deliverables

Steps	Description	Deliverables
1. Problem definition	Define: • Objectives • Deliverables • Roles and responsibilities • Current situation • Timeline • Costs and benefits	• Project plan
2. Data preparation	Prepare and become familiar with the data: • Pull together data table • Categorize the data • Clean the data • Remove unnecessary data • Transform the data • Segment the data	• High degree of familiarity with the data • Characterization of data • Documentation of the preparation steps • Data set(s) prepared for analysis
3. Implementation of the analysis	Summarizing the data • Use of summary tables, graphs, and descriptive statistics to describe the data • Use of inferential statistics to make general statements with confidence Finding hidden relationships • Identify grouping methods based on the problem • Optimize the grouping results in a controlled manner Making predictions • Select modeling approaches that match the problem and data constraints • Use separate test and training sets • Optimize model in a controlled manner	• Results of the data analysis/data mining
4. Deployment	• Plan and execute deployment based on the definition in step 1 • Measure and monitor performance • Review the project	• Deployment plan • Solution deployed • Project review

make a difference. Secondly, the quality of the data is the most important factor influencing the quality of the results. The methods used to analyze the data are not as important. Particular attention should be paid to collecting and preparing a quality data set for analysis. Thirdly, deployment is where any results obtained so far are translated into benefits to the business and this step should be carefully executed and presented to the customer in a form that they can use directly.

9.2 EXAMPLE

9.2.1 Problem Overview

To illustrate the process described in this book, we will use an example data set from Newman (1998): The Pima Indian Diabetic Database. This set is extracted from a database generated by The National Institute of Diabetes and Digestive and Kidney Diseases of the NIH. The data set contains observations on 768 female patients between age 21 and 81, and specifies whether they have contracted diabetes in five years. The following describes a hypothetical analysis scenario to illustrate the process of making sense of data.

9.2.2 Problem Definition

Objectives

Diabetes is a major cause of morbidity (for example, blindness or kidney failure) among female Pima Indians of Arizona. It is also one of the leading causes of death. The objective of the analysis is to understand any general relationships between different patient characteristics and the propensity to develop diabetes, specifically:

- **Objective 1:** Understand differences in the measurements recorded between the group that develop diabetes and the group that does not develop diabetes.
- **Objective 2:** Identify associations between the different factors and the development of diabetes that could be used for education and intervention purposes. Any associations need to make use of general categories, such as high blood pressure, to be useful.
- **Objective 3:** Develop a predictive model to estimate whether a patient will develop diabetes.

The success criterion is whether the work results in a decrease in patients developing diabetes and this result should be measured over time.

The population of this study consists of female Pima Indians between the age of 21 and 81.

Example **219**

Deliverables

The deliverables of this project include:

- **Report:** A report summarizing the data and outlining general associations that influence diabetes.

- **Prediction model software:** Software to predict patients likely to become diabetic. To be useful the models must have sensitivity and specificity values greater than 60%. The model is to be deployed over an internal network to health care professions. No explanation of the result is required; however, a degree of training would be needed for the user to help him/her understand how to interpret the results. The time to compute any prediction should be less than five minutes.

Roles and Responsibilities

A team of experts should be put together including individuals with knowledge of how the data was collected, individuals with knowledge of diabetes, along with data analysis/data mining experts and IT resources. The team should also include health care professional representatives who will eventually use the information generated. Their inclusion would both ensure their opinions are taken into account as well as to facilitate the acceptance of this new technology.

Current Situation

The team will use an available database of patient records that records whether a patient develops diabetes in five years. It is assumed that the data represents a random and unbiased sample from the population defined. The data set is available from Newman (1998).

Timeline

A timeline should be put together showing the following activities:

- **Preparation:** Assembling, characterizing, cleaning, transforming, and segmenting the data prior to analysis is essential and adequate time should be allocated for these tasks. The analysis specifically calls for an understanding of general categories and the preparation should set aside time for this preparation.

- **Implementation of the analysis:** The implementation of the analysis will involve the following data analysis/data mining tasks:

  1. **Summarizing the data:** Understanding differences in the data between the two groups will require the use of tables and graphs to summarize the data, descriptive statistics to quantify the differences, and inferential statistics to make general statements.

  2. **Finding hidden relationships:** The ability to group the data in various ways will assist in discovering unusual patterns and trends. This project

requires associations to be found within the data and the results are presented to facilitate education and prevention, that is, they need to be easy to understand. Grouping methods that satisfy these criteria should be used.

3. **Making predictions:** The project calls for the development of a prediction model. Different classification modeling approaches should be considered and optimized.

- **Deployment:** A plan for delivering the predictive model to the health care professionals over the internal network should be developed. In addition to planning the technical rollout, the appropriate training should be supplied to the health care professionals. A double blind test to monitor the deployment is not possible in this case. The monitoring of the deployment should be periodically tested against new records in the database to ensure that an appropriate level of accuracy is maintained. In addition, as the database expands, additional modeling of the data should be investigated to evaluate if results can be improved.

Costs and Benefits

A cost-benefit analysis would be useful to compare the cost of the project with the anticipated benefits of the analysis.

9.2.3   Data Preparation

Pull Together Data Table

A data set containing 768 observations has been made available. It contains patient records describing a number of attributes in addition to whether the patient went on to develop diabetes in the following five years. The data set contains the following variables:

- **Pregnant:** A record of the number of times the patient has been pregnant
- **Plasma–Glucose:** Plasma–glucose concentration measured using a two-hour oral glucose tolerance test
- **DiastolicBP:** Diastolic blood pressure
- **TricepsSFT:** Triceps skin fold thickness
- **Serum–Insulin:** Two-hour serum insulin
- **BMI:** Body mass index
- **DPF:** Diabetes pedigree function
- **Age:** Age of the patient
- **Class:** Diabetes onset within five years

Categorize Variables

Table 9.2 summarizes the variables in the data set along with their anticipated role in the analysis using the categories described in Section 3.3.

Example   **221**

Table 9.2.   Categorization of the variables in the data set

Variable	Continuous/ discrete	Scale of measurement	Anticipated role	Comments
Pregnant	Continuous	Ratio	Descriptor	Number of times pregnant
Plasma–Glucose	Continuous	Ratio	Descriptor	Plasma–glucose concentration in the blood in a two-hour oral glucose tolerance test
DiastolicBP	Continuous	Ratio	Descriptor	Units: mm Hg
TricepsSFT	Continuous	Ratio	Descriptor	Units: mm
Serum–Insulin	Continuous	Ratio	Descriptor	Units: mm U/ml
BMI	Continuous	Ratio	Descriptor	Body mass index
DPF	Continuous	Ratio	Descriptor	Diabetes pedigree function
Age	Continuous	Ratio	Descriptor	Units: years
Class	Discrete	Ordinal	Response	0 – does not contract diabetes in five years 1 – contracts diabetes in five years

Clean Variables

At this point, it is important to include those involved in collecting the data to understand how best to clean the data. A preliminary analysis of the data indicates the use of zero for missing data. Any assumptions should be validated with those involved in collection. Table 9.3 shows the number of zero values in each variable.

Removing all observations with missing data would significantly decrease the size of the data set to analyze. The variables **TricepsSFT** and **Serum–insulin** are candidates for removal and will be discussed in the next section. At this point all observations with zero values in the following variables will be removed: **Pregnant**, **Plasma–Glucose**, **DiastolicBP**, and **BMI**.

When cleaning a data set, it is useful to look for outliers in the data. For example, there is an observation with a **TricepsSFT** value of 99. This is over 6.5 standard deviations away from the mean of the data (see Figure 9.1). This

Table 9.3.   Number of zeros in each variable

Variable	Number of zero values
Pregnant	111
Plasma–Glucose	5
DiastolicBP	35
TricepsSFT	227
Serum–Insulin	374
BMI	11
DPF	0
Age	0
Class	0

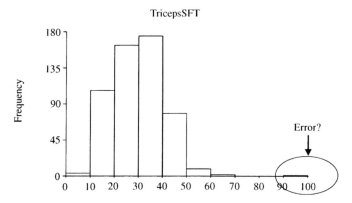

Figure 9.1.   Potential error in the data

observation should be discussed with those who collected the data to determine whether it is an error. For this analysis, the observation will be removed.

Remove Variables

An analysis of the relationship between all assigned descriptor variables was performed. There is a relationship between **BMI** and **TricepsSFT** with a value of $r = 0.67$ (see Figure 9.2). This indicates that one of these variables could be a surrogate for the other. Additionally, **TricepsSFT** has 227 missing data points. For this analysis, it will be removed based on the number of missing values and its relationship to **BMI**. The variable **Serum–Insulin** is also to be removed from the data set because of the number of missing values.

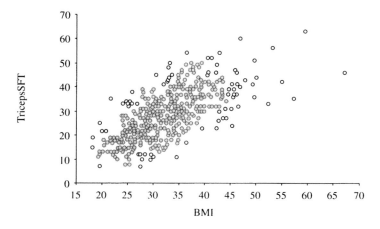

Figure 9.2.   Relationship between **TricepsSFT** and **BMI**

Example   **223**

Transform Variables

The data is further examined to determine whether any transformations of the data are required. The following transformations are considered within this analysis: normalization, discretization, and aggregation.

To ensure that all variables are considered with equal weight in any further analysis, the min-max normalization, described in Section 3.4.4, was applied to each variable where the new range is between 0 and 1.

$$Value' = \frac{Value - OriginalMin}{OriginalMax - OriginalMin}(NewMax - NewMin) + NewMin$$

Table 9.4 illustrates a portion of the new table with the newly transformed variables added to the data set.

The frequency distributions for all variables are examined to see whether the variables follow a normal distribution and therefore can be used with parametric modeling approaches without transformation. For example, the **DiastolicBP** variable follows a normal distribution and can be used without transformation (Figure 9.3).

If we wish to use the variable **Serum–insulin** within modeling approaches that require a normal distribution, the variable would require a transformation, such as a *log* transformation to satisfy this criterion. In Figure 9.4, the **Serum–insulin** variable has been applied as a *log* transformation and now reflects more closely a normal distribution.

One of the requirements of this analysis is to classify general associations between classes of variables, such as high blood pressure, and diabetes. To this end, each variable is binned into a small number of categories. This process should be performed in consultation with both any subject matter experts and/or the healthcare professionals who will use the results. This is to ensure that any subject matter or practical considerations are taken into account prior to the analysis, since the results will be presented in terms of these categories.

The following summarizes the cut-off values (shown in parentheses) along with the names of the bins for the variables:

- **Pregnant:** low (1,2), medium (3,4,5), high (> 6)
- **Plasma–Glucose:** low (< 90), medium (90–150), high (> 150)
- **DiastolicBP:** normal (< 80), normal-to-high (80–90), high (> 90)
- **BMI:** low (< 25), normal (25–30), obese (30–35), severely obese (> 35)
- **DPF:** low (< 0.4), medium (0.4–0.8), high (> 0.8)
- **Age:** 20–39, 40–59, 60 plus
- **Class:** yes (1), no (0)

Table 9.5 summarizes a sample of observations with binned values.

Aggregated variables have already been generated: **BMI** from the patient's weight and height as well as the **DPF** (diabetes pedigree function).

Table 9.4. Variables normalized to range 0 to 1

Pregnant	Pregnant (normalized)	Plasma–Glucose	Plasma–Glucose (normalized)	DiastolicBP	DiastolicBP (normalized)	BMI	BMI (normalized)	DPF	DPF (normalized)	Age	Age (normalized)
6	0.35	148	0.67	72	0.49	33.6	0.31	0.627	0.23	50	0.48
1	0.059	85	0.26	66	0.43	26.6	0.17	0.351	0.12	31	0.17
8	0.47	183	0.90	64	0.41	23.3	0.10	0.672	0.25	32	0.18
1	0.059	89	0.29	66	0.43	28.1	0.20	0.167	0.037	21	0
5	0.29	116	0.46	74	0.51	25.6	0.15	0.201	0.052	30	0.15
3	0.18	78	0.22	50	0.27	31	0.26	0.248	0.072	26	0.083
2	0.12	197	0.99	70	0.47	30.5	0.25	0.158	0.033	53	0.53
4	0.24	110	0.43	92	0.69	37.6	0.40	0.191	0.047	30	0.15
10	0.59	168	0.8	74	0.51	38	0.40	0.537	0.20	34	0.22
10	0.59	139	0.61	80	0.57	27.1	0.18	1.441	0.58	57	0.6
1	0.059	189	0.94	60	0.37	30.1	0.24	0.398	0.14	59	0.63

Example   **225**

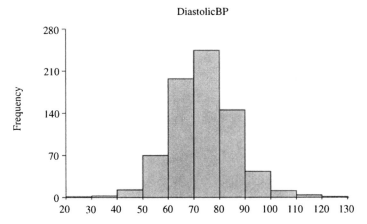

Figure 9.3.   Frequency distribution of variable **DiastolicBP**

Deliverable

The results from this stage of the project are a cleaned and transformed data set ready for analysis along with a description of the steps that were taken to create the data set. This description is useful for a number of reasons including validating the results as well as repeating the exercise later with different data. The following is a list of variables in the cleaned data table.

- **Pregnant**
- **Pregnant (grouped)**
- **Pregnant (normalized)**
- **Plasma–Glucose**
- **Plasma–Glucose (grouped)**
- **Plasma–Glucose (normalized)**
- **DiastolicBP**

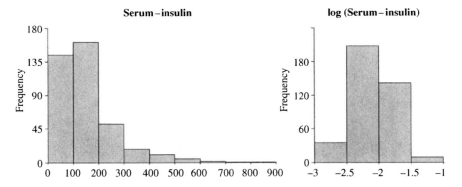

Figure 9.4.   Log transformation of **Serum–insulin**

Table 9.5. Binning of the variables

Pregnant	Pregnant (grouped)	Plasma–Glucose	Plasma–Glucose (grouped)	DiastolicBP	DiastolicBP (grouped)	BMI	BMI (grouped)	DPF	DPF (grouped)	Age	Age (grouped)	Class	Diabetes
6	high	148	medium	72	normal	33.6	obese	0.627	medium	50	40–59	1	yes
1	low	85	low	66	normal	26.6	normal	0.351	low	31	20–39	0	no
8	high	183	high	64	normal	23.3	low	0.672	medium	32	20–39	1	yes
1	low	89	low	66	normal	28.1	normal	0.167	low	21	20–39	0	no
5	medium	116	medium	74	normal	25.6	normal	0.201	low	30	20–39	0	no
3	medium	78	low	50	normal	31	obese	0.248	low	26	20–39	1	yes
2	low	197	high	70	normal	30.5	obese	0.158	low	53	40–59	1	yes
4	medium	110	medium	92	high	37.6	severely obese	0.191	low	30	20–39	0	no
10	high	168	high	74	normal	38	severely obese	0.537	medium	34	20–39	1	yes
10	high	139	medium	80	normal-to-high	27.1	normal	1.441	high	57	40–59	0	no
1	low	189	high	60	normal	30.1	obese	0.398	low	59	40–59	1	yes

Example   **227**

- **DiastolicBP (grouped)**
- **DiastolicBP (normalized)**
- **BMI**
- **BMI (grouped)**
- **BMI (normalized)**
- **DPF**
- **DPF (grouped)**
- **DPF (normalized)**
- **Age**
- **Age (grouped)**
- **Age (normalized)**
- **Class**
- **Diabetes**

In Figure 9.5, the frequency distribution of the variable class is shown. Figure 9.6 characterizes the variables assigned as descriptors. For each variable, a frequency distribution is generated and presented alongside a series of descriptive statistics in order to characterize the variables.

9.2.4   Implementation of the Analysis

Summarizing the Data

The use of graphs, summary tables, descriptive statistics, and inferential statistics will be used here to understand the differences between the two groups and the measured data. Figure 9.7 shows the distribution of the **Plasma–Glucose** variable

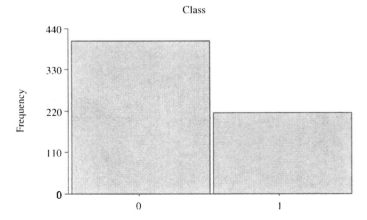

Figure 9.5.   Frequency distribution of class variables

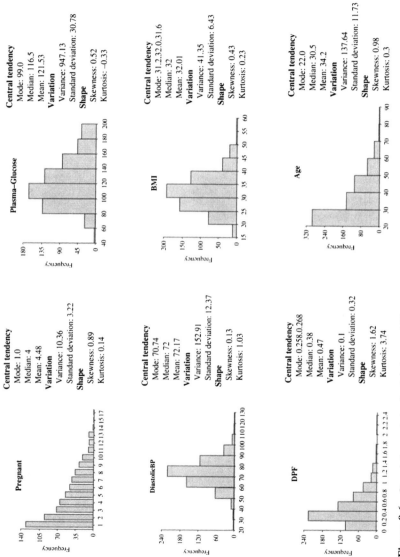

Figure 9.6. Descriptive statistics for descriptor variables

Example   **229**

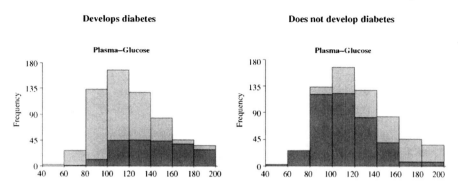

Figure 9.7.   **Plasma–Glucose** distributions for the two groups

(light gray). Observations belonging to the two groups are highlighted in dark gray. In the histogram on the left, the dark gray highlighted observations belong to the group that went on to develop diabetes. The observations highlighted on the right histogram are patients that did not develop diabetes. These graphs indicate that the distribution of **Plasma–Glucose** data between the groups is significantly different. Almost all patients with the highest **Plasma–Glucose** values went on to develop diabetes. Almost all the patients with the lowest **Plasma–Glucose** values did not go on to develop diabetes within five years. In Figure 9.8, the two groups are plotted

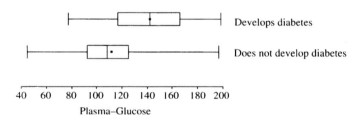

Figure 9.8.   Box plots showing **Plasma–Glucose** variation between the two groups

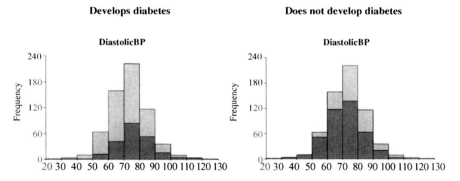

Figure 9.9.   Distribution of **DiastolicBP** between the two groups

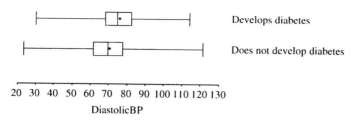

Figure 9.10.   Box plots showing **DiastolicBP** variation between the two groups

alongside each other using box plots. There is a significant shift in the central tendency of the **Plasma–Glucose** values between the two groups.

Figure 9.9 shows the distribution for the variable **DiastolicBP**. The light gray color is the overall frequency distribution and the highlighted observations on the left are the group that went on to develop diabetes. The highlighted group on the right did not develop diabetes. From these graphs it is difficult to see any discernable trends that differentiate the two groups, since the shape of the distributions is similar, even though the number of observations is higher in the group without diabetes. If we plot the two groups using a box plot, we see that the group that went on to develop diabetes is generally higher than the group that did not (Figure 9.10).

Table 9.6 summarizes the means for all variables between the group that went on to develop diabetes and the group that did not. Figure 9.11 displays the frequency distribution for all variables to understand differences between the two groups (diabetes and not diabetes). It can be seen from the graphs, that the values for **Pregnant, Plasma–Glucose, BMI** and **Age** are significantly different between the two groups. It is more difficult to see the differences between the variables **DiastolicBP** and **DPF**.

Up to this point, we have used graphs, summary tables, and descriptive statistics to visualize and characterize the differences between the two groups. We will now use inferential statistics to understand if these differences are significant enough to make claims about the general population concerning their differences. We will use a hypothesis tests to make this assessment, described in Section 5.3.3.

As an example, we will use the **DiastolicBP** variable. The observations are divided into two groups, those patients that went on to develop diabetes (group 1) and those patients that did not go on to develop diabetes (group 2). We will specify a null and alternative hypothesis:

Table 9.6.   Summary table for mean of descriptor variable for each group

Diabetes	Patient count	Mean (Pregnant)	Mean (Plasma–Glucose)	Mean (DiastolicBP)	Mean (BMI)	Mean (DPF)	Mean (Age)
no	408	3.87	110.7	70.51	30.58	0.43	31.93
yes	216	5.65	142	75.3	34.72	0.54	38.5

Example   **231**

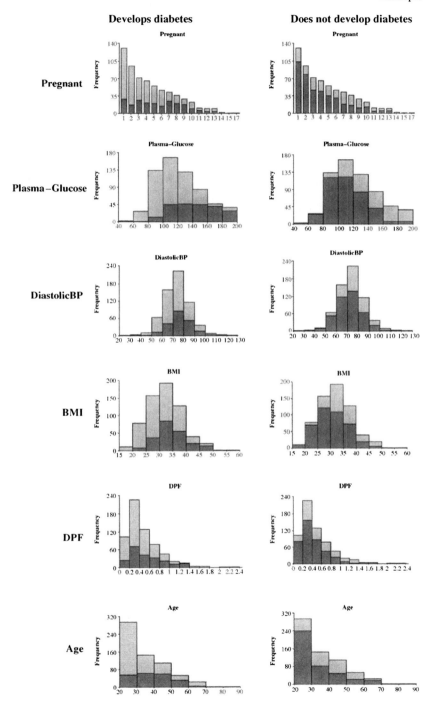

Figure 9.11. Frequency distribution for all descriptor variables across the two groups

$$H_0: \mu_1 = \mu_2$$
$$H_a: \mu_1 > \mu_2$$

where μ_1 is the population mean of group 1 and μ_2 is the population mean of group 2.

It is calculated that the sample means are $\bar{x}_1 = 75.3$ and $\bar{x}_2 = 70.51$. The number of observations in group 1 is 216 (n_1) and the number of observations in group 2 is 408 (n_2). The standard deviation of group 1 is 11.95 (s_1) with a variance of 142.8 (s_1^2) and the standard deviation of group 2 is 12.27 (s_2) with a variance of 150.6 (s_2^2). We wish to make any claims with a 99% confidence level (i.e. $\alpha = 0.01$).

The following formulas will be used to calculate the hypothesis score:

$$z = \frac{(\bar{x}_1 - \bar{x}_2) - (\mu_1 - \mu_2)}{s_P \sqrt{\frac{1}{n_1} + \frac{1}{n_2}}}$$

where

$$s_P^2 = \frac{(n_1 - 1)s_1^2 + (n_2 - 1)s_2^2}{(n_1 - 1) + (n_2 - 1)}$$

We use these formulas to calculate a hypothesis score:

$$s_P^2 = \frac{(216 - 1)142.8 + (408 - 1)150.6}{(216 - 1) + (408 - 1)} = 147.9$$

$$z = \frac{(75.3 - 70.51) - (0)}{\sqrt{147.9}\sqrt{\frac{1}{216} + \frac{1}{408}}} = 4.68$$

A *p-value* is determined, as described in Section 5.3.3, and identified using Appendix A.1. The *p-value* for this score is 0.0000014 and hence the null hypothesis is rejected and we state that there is a difference. A hypothesis score for all the variables is presented in Table 9.7.

Finding Hidden Relationships

The second objective was to identify general associations in the data to understand the relationship between the measured fields and whether the patient goes on to develop diabetes. Since the analysis will make use of categorical data, requires the identification of associations, and must be easy to interpret, the associative rule grouping approach was selected (described in Section 6.3). Using the following variables, the observations were grouped and rules extracted:

Pregnant (grouped)

Plasma–Glucose (grouped)

DiastolicBP (grouped)

Example   **233**

Table 9.7.   Hypothesis score for each variable

		count	mean	Standard deviation	Hypothesis test (z)	p-value
Pregnant	Diabetes	216	5.65	3.42	5.685519	< 0.000001
	Not Diabetes	408	3.87	3.87		
Plasma–Glucose	Diabetes	216	142	29.91	13.80435	< 0.000001
	Not Diabetes	408	110.7	25.24		
DiastolicBP	Diabetes	216	75.3	11.95	4.68117	0.0000014
	Not Diabetes	408	70.51	12.27		
BMI	Diabetes	216	34.72	6.04	8.032158	< 0.000001
	Not Diabetes	408	30.58	6.17		
DPF	Diabetes	216	0.54	0.34	4.157876	0.000017
	Not Diabetes	408	0.43	0.3		
Age	Diabetes	216	38.5	10.74	6.899296	< 0.000001
	Not Diabetes	408	31.93	11.61		

BMI (grouped)

DPF (grouped)

Age (grouped)

Diabetes

A restriction to generate groups with more than 30 observations was specified. Table 9.8 illustrates the top rules extracted from the data where the THEN-part of the rule is "Diabetes = yes". Here, all the rules describe combinations of risk factors that lead to diabetes. The rules all have high confidence values (indicating the strength of the rule) in addition to a strong positive association as indicated by the high lift scores. Table 9.9 illustrates associations where the THEN-part of the rule is "Diabetes = no". These are the highest ranking rules based on the confidence values. These rules should be discussed with the subject matter expert to determine how they should be interpreted by the health care professionals.

Table 9.8.   Associative rules in the diabetes group with highest confidence

If	Then	Support	Confidence	Lift
Plasma–Glucose (grouped) = high and Age (grouped) = 40–59	Diabetes = yes	6%	0.84	2.44
Plasma–Glucose (grouped) = high and BMI (grouped) = severely obese	Diabetes = yes	6.6%	0.82	2.37
Plasma–Glucose (grouped) = high and BMI (grouped) = obese	Diabetes = yes	5.6%	0.78	2.25
Pregnant (grouped) = high and Plasma–Glucose (grouped) = high	Diabetes = yes	7.5%	0.77	2.23

Table 9.9.   Associative rules in the not diabetes group with highest confidence

If	Then	Support	Confidence	Lift
BMI (grouped) = low and DPF (grouped) = low and Age (grouped) = 20–39	Diabetes = no	6%	1	1.53
Pregnant (grouped) = low and Plasma–Glucose (grouped) = medium and BMI (grouped) = low	Diabetes = no	5%	1	1.53
DiastolicBP (grouped) = normal and BMI (grouped) = low and DPF (grouped) = low and Age (grouped) = 20–39	Diabetes = no	5%	1	1.53
Pregnant (grouped) = low and Plasma–Glucose(grouped) = medium and DiastolicBP (grouped) = normal and DPF (grouped) = low and Age (grouped) = 20–39	Diabetes = no	8.7%	0.98	1.5
Plasma–Glucose (grouped) = medium and BMI (grouped) = low and Age (grouped) = 20–39	Diabetes = no	7.9%	0.98	1.5
Pregnant (grouped) = low and Plasma–Glucose (grouped) = low	Diabetes = no	6.6%	0.98	1.49
Pregnant (grouped) = low and BMI (grouped) = low	Diabetes = no	6.4%	0.98	1.49
Pregnant (grouped) = low and Plasma–Glucose(grouped) = low and Age (grouped) = 20–39	Diabetes = no	6.4%	0.98	1.49
Plasma–Glucose (grouped) = medium and DiastolicBP (grouped) = normal and BMI(grouped) = low and Age (grouped) = 20–39	Diabetes = no	6.4%	0.98	1.49
Pregnant (grouped) = low and Plasma–Glucose (grouped) = low and DiastolicBP (grouped) = normal	Diabetes = no	6.3%	0.98	1.49
Pregnant (grouped) = low and Plasma–Glucose (grouped) = low and DiastolicBP (grouped) = normal and Age (grouped) = 20–39	Diabetes = no	6.3%	0.98	1.49
Pregnant (grouped) = low and BMI (grouped) = low and Age (grouped) = 20–39	Diabetes = no	6.3%	0.98	1.49
Plasma–Glucose (grouped) = low and DPF (grouped) = low and Age (grouped) = 20–39	Diabetes = no	6.3%	0.98	1.49

Example   **235**

Making Predictions

The third objective of this exercise was to develop a predictive model to classify patients into two categories: (1) patients that will develop diabetes in the next five years and (2) patients that will not develop diabetes in the next five years. Since the response variable (**Class**) is categorical, we must develop a classification model. There are many alternative classification modeling approaches that we could consider. Since there is no need to explain how these results were calculated, selecting a method that generates explanations or confidence values is not necessary. We decided to select k-Nearest Neighbors (described in Section 7.3) and neural networks (described in Section 7.5) approaches to build the models. For both types of models, an experiment was designed to optimize the parameters used in generating the models. Since we are interested in both specificity and sensitivity of the results, the experiments will measure both scores. The models will be tested using a 10% cross validation (described in Section 7.1.5).

The analysis performed so far is critical to the process of developing prediction models. It helps us understand which variables are most influential, as well as helping us to interpret the results. Table 9.10 illustrates the optimization of the kNN (k-Nearest Neighbors) model using different descriptor variables with an optimal value for k, and using the Euclidean distance. The resulting model accuracy is displayed using the format "sensitivity/specificity" along with the best value of k. Table 9.11 shows a section of the optimization of the neural network models, using different input variables, different numbers of iterations, and different numbers of hidden layers. Again, the resulting model accuracy is displayed using the format "sensitivity/specificity".

The following model gave the best overall performance (both sensitivity and specificity) and was selected: neural network with two hidden layers, 50,000 cycles, and a learning rate of 0.5 using all six descriptors as inputs. The overall concordance for this model was 0.79 (or 79%) with a specificity of 0.66 (66%) and a sensitivity of 0.86 (86%).

Once the final model has been built, it is often a valuable exercise to look at observations that were not correctly predicted. Figure 9.12 presents a series of box plots for observations predicted to be in the not diabetes group, but who were diabetic (false positives). The upper box plot represents the set of false positives, the lower presents all observations. Based on our understanding of the data, diabetes is often associated with increased levels of **Plasma–Glucose**. In these examples, the patients had a lower than expected level of **Plasma–Glucose**. Other characteristics are similar to the average for the data set. This indicates that we may be missing important attributes to classify these observations correctly, such as other risk factors (e.g. level of physical activity, cholesterol level, etc.).

We can also look at examples where we predicted the patients to become diabetic when in fact they did not (false negatives). A series of box plots for the descriptor variables are presented in Figure 9.13. The upper box plots are the false negatives and the lower box plots are all observations. These patients have characteristics, based on our understanding of the data, of individuals that would go

Table 9.10. Optimization of the kNN model

						k	Sensitivity/Specificity
PRE	PG	DBP	BMI	DPF	AGE	21	0.54/0.87
-	PG	DBP	BMI	DPF	AGE	29	0.6/0.88
PRE	-	DBP	BMI	DPF	AGE	29	0.41/0.85
PRE	PG	-	BMI	DPF	AGE	21	0.56/0.88
PRE	PG	DBP	-	DPF	AGE	22	0.53/0.89
PRE	PG	DBP	BMI	-	AGE	18	0.56/0.88
PRE	PG	DBP	BMI	DPF	-	29	0.51/0.9
-	-	DBP	BMI	DPF	AGE	28	0.41/0.86
-	PG	-	BMI	DPF	AGE	27	0.62/0.87
-	PG	DBP	-	DPF	AGE	29	0.58/0.88
-	PG	DBP	BMI	-	AGE	23	0.6/0.86
-	PG	DBP	BMI	DPF	-	29	0.5/0.88
PRE	-	-	BMI	DPF	AGE	16	0.38/0.88
PRE	-	DBP	-	DPF	AGE	28	0.28/0.91
PRE	-	DBP	BMI	-	AGE	27	0.41/0.84
PRE	-	DBP	BMI	DPF	-	25	0.33/0.87
PRE	PG	-	-	DPF	AGE	28	0.51/0.89
PRE	PG	-	BMI	-	AGE	29	0.53/0.87
PRE	PG	-	BMI	DPF	-	29	0.51/0.89
PRE	PG	DBP	-	-	AGE	29	0.54/0.84
PRE	PG	DBP	-	DPF	-	28	0.49/0.9
PRE	PG	DBP	BMI	-	-	29	0.52/0.87
-	-	-	BMI	DPF	AGE	23	0.5/0.86
-	-	DBP	-	DPF	AGE	23	0.39/0.85
-	-	DBP	BMI	-	AGE	27	0.46/0.81
-	-	DBP	BMI	DPF	-	29	0.35/0.89
-	PG	-	-	DPF	AGE	23	0.58/0.86
-	PG	-	BMI	-	AGE	28	0.6/0.87
-	PG	-	BMI	DPF	-	26	0.49/0.9
-	PG	DBP	-	-	AGE	25	0.56/0.88
-	PG	DBP	-	DPF	-	29	0.51/0.88
-	PG	DBP	BMI	-	-	28	0.46/0.89
PRE	-	-	-	DPF	AGE	29	0.37/0.85
PRE	-	-	BMI	-	AGE	24	0.42/0.86
PRE	-	-	BMI	DPF	-	27	0.36/0.88
PRE	-	DBP	-	-	AGE	28	0.34/0.85
PRE	-	DBP	-	DPF	-	29	0.29/0.88
PRE	-	DBP	BMI	-	-	29	0.31/0.88
PRE	PG	-	-	-	AGE	22	0.53/0.86
PRE	PG	-	BMI	-	-	28	0.54/0.89
PRE	PG	DBP	-	-	-	29	0.48/0.87
PRE	PG	-	-	-	-	29	0.48/0.88
PRE	-	DBP	-	-	-	29	0.2/0.88
PRE	-	-	BMI	-	-	20	0.31/0.87
PRE	-	-	-	DPF	-	28	0.24/0.91

Table 9.10.   *(Continued)*

						k	Sensitivity/Specificity
PRE	-	-	-	-	AGE	29	0.39/0.81
-	PG	DBP	-	-	-	29	0.43/0.9
-	PG	-	BMI	-	-	29	0.47/0.88
-	PG	-	-	DPF	-	26	0.45/0.9
-	PG	-	-	-	AGE	28	0.52/0.87
-	-	DBP	BMI	-	-	29	0.22/0.88
-	-	DBP	-	DPF	-	29	0.19/0.91
-	-	DBP	-	-	AGE	29	0.41/0.81
-	-	-	BMI	DPF	-	29	0.36/0.87
-	-	-	BMI	-	AGE	22	0.48/0.84
-	-	-	-	DPF	AGE	28	0.39/0.85

on to develop diabetes, that is elevated **Plasma–Glucose** levels and increased **BMI**. Again, this would suggest that the data is missing important fields for the classification of this group of individuals.

9.2.5   Deployment of the Results

The deployment of the results should be carefully planned since this is how the work, put in so far, will be translated into any anticipated benefits. A report should be written by the team outlining the analysis and the results. A plan should be developed describing how the prediction model will be made available to the health care professionals including the development of any new software, as well as training the professionals to use and interpret the results. A plan for ongoing monitoring of the results and for updating the model should also be developed.

As with all projects, certain approaches worked well, whereas others did not work so well. For example, looking at the false negatives and false positives was an informative exercise. Understanding and documenting the successes and failures will allow you to share your experiences as well as improve future projects.

9.3   ADVANCED DATA MINING

9.3.1   Overview

Some common applications of exploratory data analysis and data mining require special treatment. They all can make use of the techniques described in the book; however, there are a number of factors that should be considered and the data may need to be pre-analyzed prior to using it within the framework described in the book. The further reading section of this chapter contains links to additional resources on these subjects.

Table 9.11. Optimization of the neural network model

						1 hidden layer				2 hidden layers				3 hidden layers			
						5 K	20 k	50 K	100 k	5 k	20 K	50 K	100 k	5 k	20 K	50 k	100 k
PRE	PG	DBP	BMI	DPF	AGE	0.55/0.89	0.62/0.85	0.63/0.80	0.65/0.84	0.58/0.85	0.59/0.84	0.66/0.86	0.64/0.83	0.23/.89	0.52/0.84	0.58/0.84	0.65/0.83
-	PG	DBP	BMI	DPF	AGE	0.62/0.83	0.57/0.87	0.63/0.84	0.64/0.83	0.56/0.83	0.55/0.87	0.50/0.85	0.62/0.83	0.00/1.00	0.45/0.91	0.60/0.82	0.55/0.88
PRE	-	DBP	BMI	DPF	AGE	0.28/0.89	0.45/0.83	0.45/0.83	0.48/0.79	0.23/0.91	0.33/0.87	0.40/0.83	0.46/0.81	0.00/1.00	0.49/0.79	0.49/0.77	0.44/0.79
PRE	PG	-	BMI	DPF	AGE	0.57/0.83	0.64/0.84	0.57/0.88	0.61/0.83	0.35/0.90	0.62/0.83	0.60/0.85	0.60/0.83	0.00/1.00	0.41/0.92	0.61/0.89	0.61/0.85
PRE	PG	DBP	-	DPF	AGE	0.47/0.90	0.63/0.82	0.60/0.85	0.65/0.82	0.36/0.89	0.60/0.84	0.58/0.82	0.64/0.82	0.00/1.00	0.57/0.84	0.60/0.85	0.58/0.85
PRE	PG	DBP	BMI	-	AGE	0.52/0.90	0.60/0.87	0.62/0.82	0.56/0.85	0.48/0.81	0.59/0.87	0.70/0.77	0.57/0.87	0.10/0.97	0.47/0.85	0.64/0.83	0.63/0.86

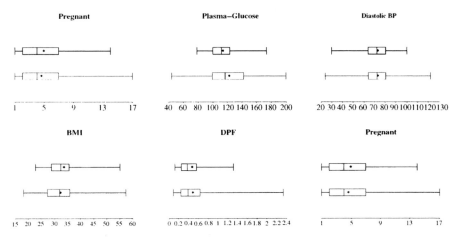

Figure 9.12. Summary of contents of false positives

9.3.2 Text Data Mining

A common application is data mining information contained in books, journals, web content, intranet content, and content on your desktop. One of the first barriers to using the data analysis and data mining techniques described in this book is the nontabular and textual format of documents. However, if the information can be translated into a tabular form then we can start to use the methods described on text documents. For example, a series of documents could be transformed into a data table as shown in Table 9.12. In this situation each row represents a different document. The columns represent all words contained in all documents. For each document, the presence of a word is indicated by "1" and the absence of a word is

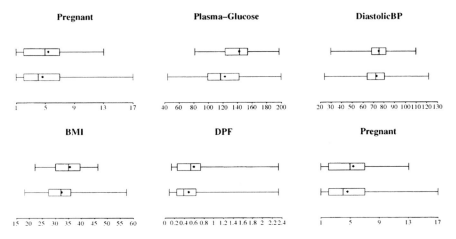

Figure 9.13. Summary of contents of false negatives

Table 9.12.   Table generated from documents

Document name	"Data'	"Mining"	"Textual"	"Analysis"
Document A	1	0	1	0
Document B	0	1	0	0
Document C	1	1	0	1
...				
Document D	0	0	0	1

indicated by "0". For example, Document A has the word "Data" somewhere in the document but no mention of the word "Mining" anywhere in the document. Once the table is in this format, the data mining approaches described can be used to group and classify documents as well as looking for word combination patterns. This is a simple view of data mining text-based unstructured documents. Additional resources are presented in the further reading section of this chapter describing methods for data mining unstructured textual documents.

9.3.3   Time Series Data Mining

In many disciplines such as financial, meteorological, and medical areas, data is collected at specific points in time. Methods for analyzing this type of data are similar to those outlined in this book. However, when looking at time series data, there are often underlying trends that need to be factored out. For example, measuring rain over the course of the year in many locations will change due to the changing of the seasons. These underlying trends need to be factored out in order to detect trends not related to seasonal variables.

9.3.4   Sequence Data Mining

In other areas events or phenomena happen in a particular sequence order, with time not being one of the dimensions to analyze. For example, web log data is comprised of sequences of pages explored. In addition to the methods described, other techniques such as hidden Markov models that make use of the state change information can be used to analyze this data.

9.4   FURTHER READING

Further information on text data mining can be found in Weiss (2004) and Berry (2003) and information on time series data mining in Last (2004).

Appendix A

Statistical Tables

A.1 NORMAL DISTRIBUTION

Table A.1 represents the area or probability (α) to the right of specific z-scores for a normal distribution (see Figure A.1). For example, the area to the right of 1.66 z-score is 0.0485.

A.2 STUDENT'S *T*-DISTRIBUTION

Critical values of t are shown in Table A.2 for various degrees of freedom (*df*). The area or probability values (α) to the right of the *t-values* (see Figure A.2) are shown in the table. For example, with 13 degrees of freedom and 0.025 probability (α) the *t-value* would be 2.160.

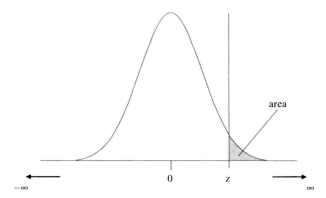

Figure A.1. Area to the right of the *z-score*

Making Sense of Data: A Practical Guide to Exploratory Data Analysis and Data Mining,
By Glenn J. Myatt
Copyright © 2007 John Wiley & Sons, Inc.

Table A.1. Standard normal distribution

z	.00	.01	.02	.03	.04	.05	.06	.07	.08	.09
0.0	0.5000	0.4960	0.4920	0.4880	0.4841	0.4801	0.4761	0.4721	0.4681	0.4641
0.1	0.4602	0.4562	0.4522	0.4483	0.4443	0.4404	0.4364	0.4325	0.4286	0.4247
0.2	0.4207	0.4168	0.4129	0.4091	0.4052	0.4013	0.3974	0.3936	0.3897	0.3859
0.3	0.3821	0.3783	0.3745	0.3707	0.3669	0.3632	0.3594	0.3557	0.3520	0.3483
0.4	0.3446	0.3409	0.3372	0.3336	0.3300	0.3264	0.3228	0.3192	0.3156	0.3121
0.5	0.3085	0.3050	0.3015	0.2981	0.2946	0.2912	0.2877	0.2843	0.2810	0.2776
0.6	0.2743	0.2709	0.2676	0.2644	0.2611	0.2579	0.2546	0.2514	0.2483	0.2451
0.7	0.2420	0.2389	0.2358	0.2327	0.2297	0.2266	0.2236	0.2207	0.2177	0.2148
0.8	0.2119	0.2090	0.2061	0.2033	0.2005	0.1977	0.1949	0.1922	0.1894	0.1867
0.9	0.1841	0.1814	0.1788	0.1762	0.1736	0.1711	0.1685	0.1660	0.1635	0.1611
1.0	0.1587	0.1563	0.1539	0.1515	0.1492	0.1469	0.1446	0.1423	0.1401	0.1379
1.1	0.1357	0.1335	0.1314	0.1292	0.1271	0.1251	0.1230	0.1210	0.1190	0.1170
1.2	0.1151	0.1131	0.1112	0.1094	0.1075	0.1057	0.1038	0.1020	0.1003	0.0985
1.3	0.0968	0.0951	0.0934	0.0918	0.0901	0.0885	0.0869	0.0853	0.0838	0.0823
1.4	0.0808	0.0793	0.0778	0.0764	0.0749	0.0735	0.0721	0.0708	0.0694	0.0681
1.5	0.0668	0.0655	0.0643	0.0630	0.0618	0.0606	0.0594	0.0582	0.0571	0.0559
1.6	0.0548	0.0537	0.0526	0.0516	0.0505	0.0495	0.0485	0.0475	0.0465	0.0455
1.7	0.0446	0.0436	0.0427	0.0418	0.0409	0.0401	0.0392	0.0384	0.0375	0.0367
1.8	0.0359	0.0351	0.0344	0.0336	0.0329	0.0322	0.0314	0.0307	0.0301	0.0294
1.9	0.0287	0.0281	0.0274	0.0268	0.0262	0.0256	0.0250	0.0244	0.0239	0.0233
2.0	0.0228	0.0222	0.0217	0.0212	0.0207	0.0202	0.0197	0.0192	0.0188	0.0183
2.1	0.0179	0.0174	0.0170	0.0166	0.0162	0.0158	0.0154	0.0150	0.0146	0.0143
2.2	0.0139	0.0136	0.0132	0.0129	0.0125	0.0122	0.0119	0.0116	0.0113	0.0110
2.3	0.0107	0.0104	0.0102	0.0099	0.0096	0.0094	0.0091	0.0089	0.0087	0.0084
2.4	0.0082	0.0080	0.0078	0.0075	0.0073	0.0071	0.0069	0.0068	0.0066	0.0064

2.5	0.0062	0.0060	0.0059	0.0057	0.0055	0.0054	0.0052	0.0051	0.0049	0.0048
2.6	0.0047	0.0045	0.0044	0.0043	0.0041	0.0040	0.0039	0.0038	0.0037	0.0036
2.7	0.0035	0.0034	0.0033	0.0032	0.0031	0.0030	0.0029	0.0028	0.0027	0.0026
2.8	0.0026	0.0025	0.0024	0.0023	0.0023	0.0022	0.0021	0.0021	0.0020	0.0019
2.9	0.0019	0.0018	0.0018	0.0017	0.0016	0.0016	0.0015	0.0015	0.0014	0.0014
3.0	0.0013	0.0013	0.0013	0.0012	0.0012	0.0011	0.0011	0.0011	0.0010	0.0010
3.1	0.00097	0.00094	0.00090	0.00087	0.00084	0.00082	0.00079	0.00076	0.00074	0.00071
3.2	0.00069	0.00066	0.00064	0.00062	0.00060	0.00058	0.00056	0.00054	0.00052	0.00050
3.3	0.00048	0.00047	0.00045	0.00043	0.00042	0.00040	0.00039	0.00038	0.00036	0.00035
3.4	0.00034	0.00032	0.00031	0.00030	0.00029	0.00028	0.00027	0.00026	0.00025	0.00024
3.5	0.00023	0.00022	0.00022	0.00021	0.00020	0.00019	0.00019	0.00018	0.00017	0.00017
3.6	0.00016	0.00015	0.00015	0.00014	0.00014	0.00013	0.00013	0.00012	0.00012	0.00011
3.7	0.00011	0.00010	0.00010	0.000096	0.000092	0.000088	0.000085	0.000082	0.000078	0.000075
3.8	0.000072	0.000069	0.000067	0.000064	0.000062	0.000059	0.000057	0.000054	0.000052	0.000050
3.9	0.000048	0.000046	0.000044	0.000042	0.000041	0.000039	0.000037	0.000036	0.000034	0.000033
4.0	0.000032	0.000030	0.000029	0.000028	0.000027	0.000026	0.000025	0.000024	0.000023	0.000022
4.1	0.000021	0.000020	0.000019	0.000018	0.000017	0.000017	0.000016	0.000015	0.000015	0.000014
4.2	0.000013	0.000013	0.000012	0.000012	0.000011	0.000011	0.000010	0.000010	0.0000093	0.0000089
4.3	0.0000085	0.0000082	0.0000078	0.0000075	0.0000071	0.0000068	0.0000065	0.0000062	0.0000059	0.0000057
4.4	0.0000054	0.0000052	0.0000049	0.0000047	0.0000045	0.0000043	0.0000041	0.0000039	0.0000037	0.0000036
4.5	0.0000034	0.0000032	0.0000031	0.0000029	0.0000028	0.0000027	0.0000026	0.0000024	0.0000023	0.0000022
4.6	0.0000021	0.0000020	0.0000019	0.0000018	0.0000017	0.0000017	0.0000016	0.0000015	0.0000014	0.0000014
4.7	0.0000013	0.0000012	0.0000011	0.0000011	0.0000011	0.0000010	0.0000010	0.00000092	0.00000088	0.00000083
4.8	0.00000079	0.00000075	0.00000072	0.00000068	0.00000065	0.00000062	0.00000059	0.00000056	0.00000053	0.00000050
4.9	0.00000048	0.00000046	0.00000043	0.00000041	0.00000039	0.00000037	0.00000035	0.00000033	0.00000032	0.00000030

Adapted with rounding from Table II of R. A. Fisher and F. Yates, *Statistical Tables for Biological, Agricultural, and Medical Research*, Sixth Edition, Pearson Education Limited, © 1963 R. A. Fisher and F. Yates.

Table A.2. Student's t-distribution

df			Upper tail area		
	0.1	0.05	0.025	0.01	0.005
1	3.078	6.314	12.706	31.821	63.657
2	1.886	2.920	4.303	6.965	9.925
3	1.638	2.353	3.182	4.541	5.841
4	1.533	2.132	2.776	3.747	4.604
5	1.476	2.015	2.571	3.365	4.032
6	1.440	1.943	2.447	3.143	3.707
7	1.415	1.895	2.365	2.998	3.499
8	1.397	1.860	2.306	2.896	3.355
9	1.383	1.833	2.262	2.821	3.250
10	1.372	1.812	2.228	2.764	3.169
11	1.363	1.796	2.201	2.718	3.106
12	1.356	1.782	2.179	2.681	3.055
13	1.350	1.771	2.160	2.650	3.012
14	1.345	1.761	2.145	2.624	2.977
15	1.341	1.753	2.131	2.602	2.947
16	1.337	1.746	2.120	2.583	2.921
17	1.333	1.740	2.110	2.567	2.898
18	1.330	1.734	2.101	2.552	2.878
19	1.328	1.729	2.093	2.539	2.861
20	1.325	1.725	2.086	2.528	2.845
21	1.323	1.721	2.080	2.518	2.831
22	1.321	1.717	2.074	2.508	2.819
23	1.319	1.714	2.069	2.500	2.807
24	1.318	1.711	2.064	2.492	2.797
25	1.316	1.708	2.060	2.485	2.787
26	1.315	1.706	2.056	2.479	2.779
27	1.314	1.703	2.052	2.473	2.771
28	1.313	1.701	2.048	2.467	2.763
29	1.311	1.699	2.045	2.462	2.756
30	1.310	1.697	2.042	2.457	2.750
40	1.303	1.684	2.021	2.423	2.704
60	1.296	1.671	2.000	2.390	2.660
120	1.289	1.658	1.980	2.358	2.617
∞	1.282	1.645	1.960	2.326	2.576

Adapted from Table III of R. A. Fisher and F. Yates, *Statistical Tables for Biological, Agricultural and Medical Research*, Sixth Edition, Pearson Education Limited, © 1963 R. A. Fisher and F. Yates.

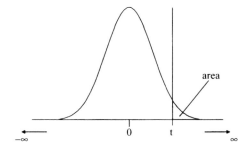

Figure A.2.   Area to the right of the *t-value*

A.3 CHI-SQUARE DISTRIBUTION

Critical values of χ^2 are shown in Table A.3 for various degrees of freedom (*df*) and illustrated in Figure A.3. The area or probability values (α) to the right of the χ^2 values are shown in the table. For example, with 9 degrees of freedom and 0.05 probability (α), the χ^2 value would be 16.919.

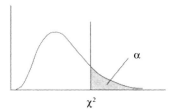

Figure A.3.   Chi-square distribution

Table A.3. χ^2 distribution

df	\multicolumn{14}{c}{Probability}
	0.99	0.98	0.95	0.90	0.80	0.70	0.50	0.30	0.20	0.10	0.05	0.02	0.01	0.001
1	0.0^3157	0.0^3628	0.00393	0.0158	0.0642	0.148	0.455	1.074	1.642	2.706	3.841	5.412	6.635	10.827
2	0.0201	0.0404	0.103	0.211	0.446	0.713	1.386	2.408	3.219	4.605	5.991	7.824	9.210	13.815
3	0.115	0.185	0.352	0.584	1.005	1.424	2.366	3.665	4.642	6.251	7.815	9.837	11.345	16.266
4	0.297	0.429	0.711	1.064	1.649	2.195	3.357	4.878	5.989	7.779	9.488	11.668	13.277	18.467
5	0.554	0.752	1.145	1.610	2.343	3.000	4.351	6.064	7.289	9.236	11.070	13.388	15.086	20.515
6	0.872	1.134	1.635	2.204	3.070	3.828	5.348	7.231	8.558	10.645	12.592	15.033	16.812	22.457
7	1.239	1.564	2.167	2.833	3.822	4.671	6.346	8.383	9.803	12.017	14.067	16.622	18.475	24.322
8	1.646	2.032	2.733	3.490	4.594	5.527	7.344	9.524	11.030	13.362	15.507	18.168	20.090	26.125
9	2.088	2.532	3.325	4.168	5.380	6.393	8.343	10.656	12.242	14.684	16.919	19.679	21.666	27.877
10	2.558	3.059	3.940	4.865	6.179	7.267	9.342	11.781	13.442	15.987	18.307	21.161	23.209	29.588
11	3.053	3.609	4.575	5.578	6.989	8.148	10.341	12.899	14.631	17.275	19.675	22.618	24.725	31.264
12	3.571	4.178	5.226	6.304	7.807	9.034	11.340	14.011	15.812	18.549	21.026	24.054	26.217	32.909
13	4.107	4.765	5.892	7.042	8.634	9.926	12.340	15.119	16.985	19.812	22.362	25.472	27.688	34.528
14	4.660	5.368	6.571	7.790	9.467	10.821	13.339	16.222	18.151	21.064	23.685	26.873	29.141	36.123
15	5.229	5.985	7.261	8.547	10.307	11.721	14.339	17.322	19.311	22.307	24.996	28.259	30.578	37.697
16	5.812	6.614	7.962	9.312	11.152	12.624	15.338	18.418	20.465	23.542	26.296	29.633	32.000	39.252
17	6.408	7.255	8.672	10.085	12.002	13.531	16.338	19.511	21.615	24.769	27.587	30.995	33.409	40.790
18	7.015	7.906	9.390	10.865	12.857	14.440	17.338	20.601	22.760	25.989	28.869	32.346	34.805	42.312
19	7.633	8.567	10.117	11.651	13.716	15.352	18.338	21.689	23.900	27.204	30.144	33.687	36.191	43.820
20	8.260	9.237	10.851	12.443	14.578	16.266	19.337	22.775	25.038	28.412	31.410	35.020	37.566	45.315

21	8.897	9.915	11.591	13.240	15.445	17.182	20.337	23.858	26.171	29.615	32.671	36.343	38.932	46.797
22	9.542	10.600	12.338	14.041	16.314	18.101	21.337	24.939	27.301	30.813	33.924	37.659	40.289	48.268
23	10.196	11.293	13.091	14.848	17.187	19.021	22.337	26.018	28.429	32.007	35.172	38.968	41.638	49.728
24	10.856	11.992	13.848	15.659	18.062	19.943	23.337	27.096	29.553	33.196	36.415	40.270	42.980	51.179
25	11.524	12.697	14.611	16.473	18.940	20.867	24.337	28.172	30.675	34.382	37.652	41.566	44.314	52.620
26	12.198	13.409	15.379	17.292	19.820	21.792	25.336	29.246	31.795	35.563	38.885	42.856	45.642	54.052
27	12.879	14.125	16.151	18.114	20.703	22.719	26.336	30.319	32.912	36.741	40.113	44.140	46.963	55.476
28	13.565	14.847	16.928	18.939	21.588	23.647	27.336	31.391	34.027	37.916	41.337	45.419	48.278	56.893
29	14.256	15.574	17.708	19.768	22.475	24.577	28.336	32.461	35.139	39.087	42.557	46.693	49.588	58.302
30	14.953	16.306	18.493	20.599	23.364	25.508	29.336	33.530	36.250	40.256	43.773	47.962	50.892	59.703
32	16.362	17.783	20.072	22.271	25.148	27.373	31.336	35.665	38.466	42.585	46.194	50.487	53.486	62.487
34	17.789	19.275	21.664	23.952	26.938	29.242	33.336	37.795	40.676	44.903	48.602	52.995	56.061	65.247
36	19.233	20.783	23.269	25.643	28.735	31.115	35.336	39.922	42.879	47.212	50.999	55.489	58.619	67.985
38	20.691	22.304	24.884	27.343	30.537	32.992	37.335	42.045	45.076	49.513	53.384	57.969	61.162	70.703
40	22.164	23.838	26.509	29.051	32.345	34.872	39.335	44.165	47.269	51.805	55.759	60.436	63.691	73.402
42	23.650	25.383	28.144	30.765	34.157	36.755	41.335	46.282	49.456	54.090	58.124	62.892	66.206	76.084
44	25.148	26.939	29.787	32.487	35.974	38.641	43.335	48.396	51.639	56.369	60.481	65.337	68.710	78.750
46	26.657	28.504	31.439	34.215	37.795	40.529	45.335	50.507	53.818	58.641	62.830	67.771	71.201	81.400
48	28.177	30.080	33.098	35.949	39.621	42.420	47.335	52.616	55.993	60.907	65.171	70.197	73.683	84.037
50	29.707	31.664	34.764	37.689	41.449	44.313	49.335	54.723	58.164	63.167	67.505	72.613	76.154	86.661
52	31.246	33.256	36.437	39.433	43.281	46.209	51.335	56.827	60.332	65.422	69.832	75.021	78.616	89.272
54	32.793	34.856	38.116	41.183	45.117	48.106	53.335	58.930	62.496	67.673	72.153	77.422	81.069	91.872
56	34.350	36.464	39.801	42.937	46.955	50.005	55.335	61.031	64.658	69.919	74.468	79.815	83.513	94.461
58	35.913	38.078	41.492	44.696	48.797	51.906	57.335	63.129	66.816	72.160	76.778	82.201	85.950	97.039
60	37.485	39.699	43.188	46.459	50.641	53.809	59.335	65.227	68.972	74.397	79.082	84.580	88.379	99.607

Table A.3. *(Continued)*

df	0.99	0.98	0.95	0.90	0.80	0.70	0.50	0.30	0.20	0.10	0.05	0.02	0.01	0.001
							Probability							
62	39.063	41.327	44.889	48.226	52.487	55.714	61.335	67.322	71.125	76.630	81.381	86.953	90.802	102.166
64	40.649	42.960	46.595	49.996	54.336	57.620	63.335	69.416	73.276	78.860	83.675	89.320	93.217	104.716
66	42.240	44.599	48.305	51.770	56.188	59.527	65.335	71.508	75.424	81.085	85.965	91.681	95.626	107.258
68	43.838	46.244	50.020	53.548	58.042	61.436	67.335	73.600	77.571	83.308	88.250	94.037	98.028	109.791
70	45.442	47.893	51.739	55.329	59.898	63.346	69.334	75.689	79.715	85.527	90.531	96.388	100.425	112.317

Adapted from Table IV of R. A. Fisher and F. Yates, *Statistical Tables for Biological, Agricultural and Medical Research*, Sixth Edition, Pearson Education Limited, © 1963 R. A. Fisher and F. Yates.

A.4 F-DISTRIBUTION

Tables A.4, A.5, A.6 and A.7 show the *F-statistics* for four different values of α: 0.1, 0.05, 0.01 and 0.005. v_1 is the number of degrees of freedom for the numerator and v_2 is the number of degrees of freedom for the denominator. Figure A.4 illustrates the F-distribution. For example, to look up a critical value for the F-statistics where the numerator degrees of freedom (v_1) are 6 and the denominator degrees of freedom (v_2) are 15 and α is 0.05, using Table A.5, is 3.94.

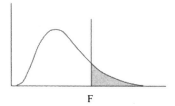

F

Figure A.4. F-distribution

Table A.4. F-distribution where $\alpha = 0.1$

V1 \ V2	1	2	3	4	5	6	7	8	9	10	12	15	20	24	30	40	60	120	∞
1	39.86	49.50	53.59	55.83	57.24	58.20	58.91	59.44	59.86	60.19	60.71	61.22	61.74	62.00	62.26	62.53	62.79	63.06	63.33
2	8.53	9.00	9.16	9.24	9.29	9.33	9.35	9.37	9.38	9.39	9.41	9.42	9.44	9.45	9.46	9.47	9.47	9.48	9.49
3	5.54	5.46	5.39	5.34	5.31	5.28	5.27	5.25	5.24	5.23	5.22	5.20	5.18	5.18	5.17	5.16	5.15	5.14	5.13
4	4.54	4.32	4.19	4.11	4.05	4.01	3.98	3.95	3.94	3.92	3.90	3.87	3.84	3.83	3.82	3.80	3.79	3.78	3.76
5	4.06	3.78	3.62	3.52	3.45	3.40	3.37	3.34	3.32	3.30	3.27	3.24	3.21	3.19	3.17	3.16	3.14	3.12	3.10
6	3.78	3.46	3.29	3.18	3.11	3.05	3.01	2.98	2.96	2.94	2.90	2.87	2.84	2.82	2.80	2.78	2.76	2.74	2.72
7	3.59	3.26	3.07	2.96	2.88	2.83	2.78	2.75	2.72	2.70	2.67	2.63	2.59	2.58	2.56	2.54	2.51	2.49	2.47
8	3.46	3.11	2.92	2.81	2.73	2.67	2.62	2.59	2.56	2.54	2.50	2.46	2.42	2.40	2.38	2.36	2.34	2.32	2.29
9	3.36	3.01	2.81	2.69	2.61	2.55	2.51	2.47	2.44	2.42	2.38	2.34	2.30	2.28	2.25	2.23	2.21	2.18	2.16
10	3.29	2.92	2.73	2.61	2.52	2.46	2.41	2.38	2.35	2.32	2.28	2.24	2.20	2.18	2.16	2.13	2.11	2.08	2.06
11	3.23	2.86	2.66	2.54	2.45	2.39	2.34	2.30	2.27	2.25	2.21	2.17	2.12	2.10	2.08	2.05	2.03	2.00	1.97
12	3.18	2.81	2.61	2.48	2.39	2.33	2.28	2.24	2.21	2.19	2.15	2.10	2.06	2.04	2.01	1.99	1.96	1.93	1.90
13	3.14	2.76	2.56	2.43	2.35	2.28	2.23	2.20	2.16	2.14	2.10	2.05	2.01	1.98	1.96	1.93	1.90	1.88	1.85
14	3.10	2.73	2.52	2.39	2.31	2.24	2.19	2.15	2.12	2.10	2.05	2.01	1.96	1.94	1.91	1.89	1.86	1.83	1.80
15	3.07	2.70	2.49	2.36	2.27	2.21	2.16	2.12	2.09	2.06	2.02	1.97	1.92	1.90	1.87	1.85	1.82	1.79	1.76
16	3.05	2.67	2.46	2.33	2.24	2.18	2.13	2.09	2.06	2.03	1.99	1.94	1.89	1.87	1.84	1.81	1.78	1.75	1.72
17	3.03	2.64	2.44	2.31	2.22	2.15	2.10	2.06	2.03	2.00	1.96	1.91	1.86	1.84	1.81	1.78	1.75	1.72	1.69
18	3.01	2.62	2.42	2.29	2.20	2.13	2.08	2.04	2.00	1.98	1.93	1.89	1.84	1.81	1.78	1.75	1.72	1.69	1.66
19	2.99	2.61	2.40	2.27	2.18	2.11	2.06	2.02	1.98	1.96	1.91	1.86	1.81	1.79	1.76	1.73	1.70	1.67	1.63

20	2.97	2.59	2.38	2.25	2.16	2.09	2.04	2.00	1.96	1.94	1.89	1.84	1.79	1.77	1.74	1.71	1.68	1.64	1.61
21	2.96	2.57	2.36	2.23	2.14	2.08	2.02	1.98	1.95	1.92	1.87	1.83	1.78	1.75	1.72	1.69	1.66	1.62	1.59
22	2.95	2.56	2.35	2.22	2.13	2.06	2.01	1.97	1.93	1.90	1.86	1.81	1.76	1.73	1.70	1.67	1.64	1.60	1.57
23	2.94	2.55	2.34	2.21	2.11	2.05	1.99	1.95	1.92	1.89	1.84	1.80	1.74	1.72	1.69	1.66	1.62	1.59	1.55
24	2.93	2.54	2.33	2.19	2.10	2.04	1.98	1.94	1.91	1.88	1.83	1.78	1.73	1.70	1.67	1.64	1.61	1.57	1.53
25	2.92	2.53	2.32	2.18	2.09	2.02	1.97	1.93	1.89	1.87	1.82	1.77	1.72	1.69	1.66	1.63	1.59	1.56	1.52
26	2.91	2.52	2.31	2.17	2.08	2.01	1.96	1.92	1.88	1.86	1.81	1.76	1.71	1.68	1.65	1.61	1.58	1.54	1.50
27	2.90	2.51	2.30	2.17	2.07	2.00	1.95	1.91	1.87	1.85	1.80	1.75	1.70	1.67	1.64	1.60	1.57	1.53	1.49
28	2.89	2.50	2.29	2.16	2.06	2.00	1.94	1.90	1.87	1.84	1.79	1.74	1.69	1.66	1.63	1.59	1.56	1.52	1.48
29	2.89	2.50	2.28	2.15	2.06	1.99	1.93	1.89	1.86	1.83	1.78	1.73	1.68	1.65	1.62	1.58	1.55	1.51	1.47
30	2.88	2.49	2.28	2.14	2.05	1.98	1.93	1.88	1.85	1.82	1.77	1.72	1.67	1.64	1.61	1.57	1.54	1.50	1.46
40	2.84	2.44	2.23	2.09	2.00	1.93	1.87	1.83	1.79	1.76	1.71	1.66	1.61	1.57	1.54	1.51	1.47	1.42	1.38
60	2.79	2.39	2.18	2.04	1.95	1.87	1.82	1.77	1.74	1.71	1.66	1.60	1.54	1.51	1.48	1.44	1.40	1.35	1.29
120	2.75	2.35	2.13	1.99	1.90	1.82	1.77	1.72	1.68	1.65	1.60	1.55	1.48	1.45	1.41	1.37	1.32	1.26	1.19
∞	2.71	2.30	2.08	1.94	1.85	1.77	1.72	1.67	1.63	1.60	1.55	1.49	1.42	1.38	1.34	1.30	1.24	1.17	1.00

Table A.5. F-distribution where $\alpha = 0.05$

v_1 \ v_2	1	2	3	4	5	6	7	8	9	10	12	15	20	24	30	40	60	120	∞
1	161.4	199.5	215.7	224.6	230.2	234.0	236.8	238.9	240.5	241.9	243.9	245.9	248.0	249.1	250.1	251.1	252.2	253.3	254.3
2	18.51	19.00	19.16	19.25	19.30	19.33	19.35	19.37	19.38	19.40	19.41	19.43	19.45	19.45	19.46	19.47	19.48	19.49	19.50
3	10.13	9.55	9.28	9.12	9.01	8.94	8.89	8.85	8.81	8.79	8.74	8.70	8.66	8.64	8.62	8.59	8.57	8.55	8.53
4	7.71	6.94	6.59	6.39	6.26	6.16	6.09	6.04	6.00	5.96	5.91	5.86	5.80	5.77	5.75	5.72	5.69	5.66	5.63
5	6.61	5.79	5.41	5.19	5.05	4.95	4.88	4.82	4.77	4.74	4.68	4.62	4.56	4.53	4.50	4.46	4.43	4.40	4.36
6	5.99	5.14	4.76	4.53	4.39	4.28	4.21	4.15	4.10	4.06	4.00	3.94	3.87	3.84	3.81	3.77	3.74	3.70	3.67
7	5.59	4.74	4.35	4.12	3.97	3.87	3.79	3.73	3.68	3.64	3.57	3.51	3.44	3.41	3.38	3.34	3.30	3.27	3.23
8	5.32	4.46	4.07	3.84	3.69	3.58	3.50	3.44	3.39	3.35	3.28	3.22	3.15	3.12	3.08	3.04	3.01	2.97	2.93
9	5.12	4.26	3.86	3.63	3.48	3.37	3.29	3.23	3.18	3.14	3.07	3.01	2.94	2.90	2.86	2.83	2.79	2.75	2.71
10	4.96	4.10	3.71	3.48	3.33	3.22	3.14	3.07	3.02	2.98	2.91	2.85	2.77	2.74	2.70	2.66	2.62	2.58	2.54
11	4.84	3.98	3.59	3.36	3.20	3.09	3.01	2.95	2.90	2.85	2.79	2.72	2.65	2.61	2.57	2.53	2.49	2.45	2.40
12	4.75	3.89	3.49	3.26	3.11	3.00	2.91	2.85	2.80	2.75	2.69	2.62	2.54	2.51	2.47	2.43	2.38	2.34	2.30
13	4.67	3.81	3.41	3.18	3.03	2.92	2.83	2.77	2.71	2.67	2.60	2.53	2.46	2.42	2.38	2.34	2.30	2.25	2.21
14	4.60	3.74	3.34	3.11	2.96	2.85	2.76	2.70	2.65	2.60	2.53	2.46	2.39	2.35	2.31	2.27	2.22	2.18	2.13
15	4.54	3.68	3.29	3.06	2.90	2.79	2.71	2.64	2.59	2.54	2.48	2.40	2.33	2.29	2.25	2.20	2.16	2.11	2.07
16	4.49	3.63	3.24	3.01	2.85	2.74	2.66	2.59	2.54	2.49	2.42	2.35	2.28	2.24	2.19	2.15	2.11	2.06	2.01
17	4.45	3.59	3.20	2.96	2.81	2.70	2.61	2.55	2.49	2.45	2.38	2.31	2.23	2.19	2.15	2.10	2.06	2.01	1.96
18	4.41	3.55	3.16	2.93	2.77	2.66	2.58	2.51	2.46	2.41	2.34	2.27	2.19	2.15	2.11	2.06	2.02	1.97	1.92
19	4.38	3.52	3.13	2.90	2.74	2.63	2.54	2.48	2.42	2.38	2.31	2.23	2.16	2.11	2.07	2.03	1.98	1.93	1.88

20	4.35	3.49	3.10	2.87	2.71	2.60	2.51	2.45	2.39	2.35	2.28	2.20	2.12	2.08	2.04	1.99	1.95	1.90	1.84
21	4.32	3.47	3.07	2.84	2.68	2.57	2.49	2.42	2.37	2.32	2.25	2.18	2.10	2.05	2.01	1.96	1.92	1.87	1.81
22	4.30	3.44	3.05	2.82	2.66	2.55	2.46	2.40	2.34	2.30	2.23	2.15	2.07	2.03	1.98	1.94	1.89	1.84	1.78
23	4.28	3.42	3.03	2.80	2.64	2.53	2.44	2.37	2.32	2.27	2.20	2.13	2.05	2.01	1.96	1.91	1.86	1.81	1.76
24	4.26	3.40	3.01	2.78	2.62	2.51	2.42	2.36	2.30	2.25	2.18	2.11	2.03	1.98	1.94	1.89	1.84	1.79	1.73
25	4.24	3.39	2.99	2.76	2.60	2.49	2.40	2.34	2.28	2.24	2.16	2.09	2.01	1.96	1.92	1.87	1.82	1.77	1.71
26	4.23	3.37	2.98	2.74	2.59	2.47	2.39	2.32	2.27	2.22	2.15	2.07	1.99	1.95	1.90	1.85	1.80	1.75	1.69
27	4.21	3.35	2.96	2.73	2.57	2.46	2.37	2.31	2.25	2.20	2.13	2.06	1.97	1.93	1.88	1.84	1.79	1.73	1.67
28	4.20	3.34	2.95	2.71	2.56	2.45	2.36	2.29	2.24	2.19	2.12	2.04	1.96	1.91	1.87	1.82	1.77	1.71	1.65
29	4.18	3.33	2.93	2.70	2.55	2.43	2.35	2.28	2.22	2.18	2.10	2.03	1.94	1.90	1.85	1.81	1.75	1.70	1.64
30	4.17	3.32	2.92	2.69	2.53	2.42	2.33	2.27	2.21	2.16	2.09	2.01	1.93	1.89	1.84	1.79	1.74	1.68	1.62
40	4.08	3.23	2.84	2.61	2.45	2.34	2.25	2.18	2.12	2.08	2.00	1.92	1.84	1.79	1.74	1.69	1.64	1.58	1.51
60	4.00	3.15	2.76	2.53	2.37	2.25	2.17	2.10	2.04	1.99	1.92	1.84	1.75	1.70	1.65	1.59	1.53	1.47	1.39
120	3.92	3.07	2.68	2.45	2.29	2.17	2.09	2.02	1.96	1.91	1.83	1.75	1.66	1.61	1.55	1.50	1.43	1.35	1.25
∞	3.84	3.00	2.60	2.37	2.21	2.10	2.01	1.94	1.88	1.83	1.75	1.67	1.57	1.52	1.46	1.39	1.32	1.22	1.00

Table A.6. F-distribution where $\alpha = 0.01$

V_1 \ V_2	1	2	3	4	5	6	7	8	9	10	12	15	20	24	30	40	60	120	∞
1	4052	4999.5	5403	5625	5764	5859	5928	5982	6022	6056	6106	6157	6209	6235	6261	6287	6313	6339	6366
2	98.50	99.00	99.17	99.25	99.30	99.33	99.36	99.37	99.39	99.40	99.42	99.43	99.45	99.46	99.47	99.47	99.48	99.49	99.50
3	34.12	30.82	29.46	28.71	28.24	27.91	27.67	27.49	27.35	27.23	27.05	26.87	26.69	26.60	26.50	26.41	26.32	26.22	26.13
4	21.20	18.00	16.69	15.98	15.52	15.21	14.98	14.80	14.66	14.55	14.37	14.20	14.02	13.93	13.84	13.75	13.65	13.56	13.46
5	16.26	13.27	12.06	11.39	10.97	10.67	10.46	10.29	10.16	10.05	9.89	9.72	9.55	9.47	9.38	9.29	9.20	9.11	9.02
6	13.75	10.92	9.78	9.15	8.75	8.47	8.26	8.10	7.98	7.87	7.72	7.56	7.40	7.31	7.23	7.14	7.06	6.97	6.88
7	12.25	9.55	8.45	7.85	7.46	7.19	6.99	6.84	6.72	6.62	6.47	6.31	6.16	6.07	5.99	5.91	5.82	5.74	5.65
8	11.26	8.65	7.59	7.01	6.63	6.37	6.18	6.03	5.91	5.81	5.67	5.52	5.36	5.28	5.20	5.12	5.03	4.95	4.86
9	10.56	8.02	6.99	6.42	6.06	5.80	5.61	5.47	5.35	5.26	5.11	4.96	4.81	4.73	4.65	4.57	4.48	4.40	4.31
10	10.04	7.56	6.55	5.99	5.64	5.39	5.20	5.06	4.94	4.85	4.71	4.56	4.41	4.33	4.25	4.17	4.08	4.00	3.91
11	9.65	7.21	6.22	5.67	5.32	5.07	4.89	4.74	4.63	4.54	4.40	4.25	4.10	4.02	3.94	3.86	3.78	3.69	3.60
12	9.33	6.93	5.95	5.41	5.06	4.82	4.64	4.50	4.39	4.30	4.16	4.01	3.86	3.78	3.70	3.62	3.54	3.45	3.36
13	9.07	6.70	5.74	5.21	4.86	4.62	4.44	4.30	4.19	4.10	3.96	3.82	3.66	3.59	3.51	3.43	3.34	3.25	3.17
14	8.86	6.51	5.56	5.04	4.69	4.46	4.28	4.14	4.03	3.94	3.80	3.66	3.51	3.43	3.35	3.27	3.18	3.09	3.00
15	8.68	6.36	5.42	4.89	4.56	4.32	4.14	4.00	3.89	3.80	3.67	3.52	3.37	3.29	3.21	3.13	3.05	2.96	2.87
16	8.53	6.23	5.29	4.77	4.44	4.20	4.03	3.89	3.78	3.69	3.55	3.41	3.26	3.18	3.10	3.02	2.93	2.84	2.75
17	8.40	6.11	5.18	4.67	4.34	4.10	3.93	3.79	3.68	3.59	3.46	3.31	3.16	3.08	3.00	2.92	2.83	2.75	2.65
18	8.29	6.01	5.09	4.58	4.25	4.01	3.84	3.71	3.60	3.51	3.37	3.23	3.08	3.00	2.92	2.84	2.75	2.66	2.57
19	8.18	5.93	5.01	4.50	4.17	3.94	3.77	3.63	3.52	3.43	3.30	3.15	3.00	2.92	2.84	2.76	2.67	2.58	2.49

20	8.10	5.85	4.94	4.43	4.10	3.87	3.70	3.56	3.46	3.37	3.23	3.09	2.94	2.86	2.78	2.69	2.61	2.52	2.42
21	8.02	5.78	4.87	4.37	4.04	3.81	3.64	3.51	3.40	3.31	3.17	3.03	2.88	2.80	2.72	2.64	2.55	2.46	2.36
22	7.95	5.72	4.82	4.31	3.99	3.76	3.59	3.45	3.35	3.26	3.12	2.98	2.83	2.75	2.67	2.58	2.50	2.40	2.31
23	7.88	5.66	4.76	4.26	3.94	3.71	3.54	3.41	3.30	3.21	3.07	2.93	2.78	2.70	2.62	2.54	2.45	2.35	2.26
24	7.82	5.61	4.72	4.22	3.90	3.67	3.50	3.36	3.26	3.17	3.03	2.89	2.74	2.66	2.58	2.49	2.40	2.31	2.21
25	7.77	5.57	4.68	4.18	3.85	3.63	3.46	3.32	3.22	3.13	2.99	2.85	2.70	2.62	2.54	2.45	2.36	2.27	2.17
26	7.72	5.53	6.64	4.14	3.82	3.59	3.42	3.29	3.18	3.09	2.96	2.81	2.66	2.58	2.50	2.42	2.33	2.23	2.13
27	7.68	5.49	4.60	4.11	3.78	3.56	3.39	3.26	3.15	3.06	2.93	2.78	2.63	2.55	2.47	2.38	2.29	2.20	2.10
28	7.64	5.45	4.57	4.07	3.75	3.53	3.36	3.23	3.12	3.03	2.90	2.75	2.60	2.52	2.44	2.35	2.26	2.17	2.06
29	7.60	5.42	4.54	4.04	3.73	3.50	3.33	3.20	3.09	3.00	2.87	2.73	2.57	2.49	2.41	2.33	2.23	2.14	2.03
30	7.56	5.39	4.51	4.02	3.70	3.47	3.30	3.17	3.07	2.98	2.84	2.70	2.55	2.47	2.39	2.30	2.21	2.11	2.01
40	7.31	5.18	4.31	3.83	3.51	3.29	3.12	2.99	2.89	2.80	2.66	2.52	2.37	2.29	2.20	2.11	2.02	1.92	1.80
60	7.08	4.98	4.13	3.65	3.34	3.12	2.95	2.82	2.72	2.63	2.50	2.35	2.20	2.12	2.03	1.94	1.84	1.73	1.60
120	6.85	4.79	3.95	3.48	3.17	2.96	2.79	2.66	2.56	2.47	2.34	2.19	2.03	1.95	1.86	1.76	1.66	1.53	1.38
∞	6.63	4.61	3.78	3.32	3.02	2.80	2.64	2.51	2.41	2.32	2.18	2.04	1.88	1.79	1.70	1.59	1.47	1.32	1.00

Adapted from E. S. Pearson and H. O. Hartley, *Biometrika Tables for Statisticians*, Vol. 1, 1958, pp. 157–63, Table 18, by permission of the Biometrika Trustees.

Table A.7. F-distribution where $\alpha = 0.005$

V_1 \ V_2	1	2	3	4	5	6	7	8	9	10	12	15	20	24	30	40	60	120	∞
1	16211	20000	21615	22500	23056	23437	23715	23925	24091	24224	24426	24630	24836	24940	25044	25148	25253	25359	25465
2	198.5	199.0	199.2	199.2	199.3	199.3	199.4	199.4	199.4	199.4	199.4	199.4	199.4	199.5	199.5	199.5	199.5	199.5	199.5
3	55.55	49.80	47.47	46.19	45.39	44.84	44.43	44.13	43.88	43.69	43.39	43.08	42.78	42.62	42.47	42.31	42.15	41.99	41.83
4	31.33	26.28	24.26	23.15	22.46	21.97	21.62	21.35	21.14	20.97	20.70	20.44	20.17	20.03	19.89	19.75	19.61	19.47	19.32
5	22.78	18.31	16.53	15.56	14.96	14.51	14.20	13.96	13.77	13.62	13.38	13.15	12.90	12.78	12.66	12.53	12.40	12.27	12.14
6	18.63	14.54	12.92	12.03	11.46	11.07	10.79	10.57	10.39	10.25	10.03	9.81	9.59	9.47	9.36	9.24	9.12	9.00	8.88
7	16.24	12.40	10.88	10.05	9.52	9.16	8.89	8.68	8.51	8.38	8.18	7.97	7.75	7.65	7.53	7.42	7.31	7.19	7.08
8	14.69	11.04	9.60	8.81	8.30	7.95	7.69	7.50	7.34	7.21	7.01	6.81	6.61	6.50	6.40	6.29	6.18	6.06	5.95
9	13.61	10.11	8.72	7.96	7.47	7.13	6.88	6.69	6.54	6.42	6.23	6.03	5.83	5.73	5.62	5.52	5.41	5.30	5.19
10	12.83	9.43	8.08	7.34	6.87	6.54	6.30	6.12	5.97	5.85	5.66	5.47	5.27	5.17	5.07	4.97	4.86	4.75	4.64
11	12.23	8.91	7.60	6.88	6.42	6.10	5.86	5.68	5.54	5.42	5.24	5.05	4.86	4.76	4.65	4.55	4.44	4.34	4.23
12	11.75	8.51	7.23	6.52	6.07	5.76	5.52	5.35	5.20	5.09	4.91	4.72	4.53	4.43	4.33	4.23	4.12	4.01	3.90
13	11.37	8.19	6.93	6.23	5.79	5.48	5.25	5.08	4.94	4.82	4.64	4.46	4.27	4.17	4.07	3.97	3.87	3.76	3.65
14	11.06	7.92	6.68	6.00	5.56	5.26	5.03	4.86	4.72	4.60	4.43	4.25	4.06	3.96	3.86	3.76	3.66	3.55	3.44
15	10.80	7.70	6.48	5.80	5.37	5.07	4.85	4.67	4.54	4.42	4.25	4.07	3.88	3.79	3.69	3.58	3.48	3.37	3.26
16	10.58	7.51	6.30	5.64	5.21	4.91	4.69	4.52	4.38	4.27	4.10	3.92	3.73	3.64	3.54	3.44	3.33	3.22	3.11
17	10.38	7.35	6.16	5.50	5.07	4.78	4.56	4.39	4.25	4.14	3.97	3.79	3.61	3.51	3.41	3.31	3.21	3.10	2.98
18	10.22	7.21	6.03	5.37	4.96	4.66	4.44	4.28	4.14	4.03	3.86	3.68	3.50	3.40	3.30	3.20	3.10	2.99	2.87
19	10.07	7.09	5.92	5.27	4.85	4.56	4.34	4.18	4.04	3.93	3.76	3.59	3.40	3.31	3.21	3.11	3.00	2.89	2.78

20	9.94	6.99	5.82	5.17	4.76	4.47	4.26	4.09	3.96	3.85	3.68	3.50	3.32	3.22	3.12	3.02	2.92	2.81	2.69
21	9.83	6.89	5.73	5.09	4.68	4.39	4.18	4.01	3.88	3.77	3.60	3.43	3.24	3.15	3.05	2.95	2.84	2.73	2.61
22	9.73	6.81	5.65	5.02	4.61	4.32	4.11	3.94	3.81	3.70	3.54	3.36	3.18	3.08	2.98	2.88	2.77	2.66	2.55
23	9.63	6.73	5.58	4.95	4.54	4.26	4.05	3.88	3.75	3.64	3.47	3.30	3.12	3.02	2.92	2.82	2.71	2.60	2.48
24	9.55	6.66	5.52	4.89	4.49	4.20	3.99	3.83	3.69	3.59	3.42	3.25	3.06	2.97	2.87	2.77	2.66	2.55	2.43
25	9.48	6.60	5.46	4.84	4.43	4.15	3.94	3.78	3.64	3.54	3.37	3.20	3.01	2.92	2.82	2.72	2.61	2.50	2.38
26	9.41	6.54	5.41	4.79	4.38	4.10	3.89	3.73	3.60	3.49	3.33	3.15	2.97	2.87	2.77	2.67	2.56	2.45	2.33
27	9.34	6.49	5.36	4.74	4.34	4.06	3.85	3.69	3.56	3.45	3.28	3.11	2.93	2.83	2.73	2.63	2.52	2.41	2.29
28	9.28	6.44	5.32	4.70	4.30	4.02	3.81	3.65	3.52	3.41	3.25	3.07	2.89	2.79	2.69	2.59	2.48	2.37	2.25
29	9.23	6.40	5.28	4.66	4.26	3.98	3.77	3.61	3.48	3.38	3.21	3.04	2.86	2.76	2.66	2.56	2.45	2.33	2.21
30	9.18	6.35	5.24	4.62	4.23	3.95	3.74	3.58	3.45	3.34	3.18	3.01	2.82	2.73	2.63	2.52	2.42	2.30	2.18
40	8.83	6.07	4.98	4.37	3.99	3.71	3.51	3.35	3.22	3.12	2.95	2.78	2.60	2.50	2.40	2.30	2.18	2.06	1.93
60	8.49	5.79	4.73	4.14	3.76	3.49	3.29	3.13	3.01	2.90	2.74	2.57	2.39	2.29	2.19	2.08	1.96	1.83	1.69
120	8.18	5.54	4.50	3.92	3.55	3.28	3.09	2.93	2.81	2.71	2.54	2.37	2.19	2.09	1.98	1.87	1.75	1.61	1.43
∞	7.88	5.30	4.28	3.72	3.35	3.09	2.90	2.74	2.62	2.52	2.36	2.19	2.00	1.90	1.79	1.67	1.53	1.36	1.00

Adapted from E. S. Pearson and H. O. Hartley, *Biometrika Tables for Statisticians*, Vol. 1, 1958, pp. 157–63. Table 18, by permission of the Biometrika Trustees.

Appendix B

Answers to Exercises

Chapter 3

   1a. Discrete

   1b. Continuous

   1c. Dichotomous

   1d. Discrete

   1e. Continuous

   1f. Continuous

   1g. Continuous

   1h. Continuous

   1i. Continuous

   1j. Binary

   2a. Nominal

   2b. Ratio

   2c. Nominal

   2d. Nominal

   2e. Ratio

   2f. Ratio

   2g. Ratio

   2h. Ratio

   2i. Ratio

   2j. Nominal

   3a. **Name**

   3b. **Age, Weight, Systolic blood pressure, Diastolic blood pressure**

   3c. **Diabetes**

   4. See Table B.1

Making Sense of Data: A Practical Guide to Exploratory Data Analysis and Data Mining,
By Glenn J. Myatt
Copyright © 2007 John Wiley & Sons, Inc.

5. See Table B.2
6. See Table B.3
7. See Table B.4 and Table B.5

Table B.1.   Chapter 3, question 4 answer

Name	Weight (kg)	Weight (kg) normalized to 0–1
P. Lee	50	0.095
R. Jones	115	0.779
J. Smith	96	0.579
A. Patel	41	0
M. Owen	79	0.4
S. Green	109	0.716
N. Cook	73	0.337
W. Hands	104	0.663
P. Rice	64	0.242
F. Marsh	136	1

Table B.2.   Chapter 3, question 5 answer

Name	Weight (kg)	Weight (kg) categorized [low, medium, high]
P. Lee	50	low
R. Jones	115	high
J. Smith	96	medium
A. Patel	41	low
M. Owen	79	medium
S. Green	109	high
N. Cook	73	medium
W. Hands	104	high
P. Rice	64	medium
F. Marsh	136	high

Table B.3.   Chapter 3, question 6 answer

Name	Weight (kg)	Height (m)	BMI
P. Lee	50	1.52	21.6
R. Jones	115	1.77	36.7
J. Smith	96	1.83	28.7
A. Patel	41	1.55	17.1
M. Owen	79	1.82	23.8
S. Green	109	1.89	30.5
N. Cook	73	1.76	23.6
W. Hands	104	1.71	35.6
P. Rice	64	1.74	21.1
F. Marsh	136	1.78	42.9

Table B.4. Chapter 3, question 7 answer (female patients)

Name	Age	Gender	Blood group	Weight (kg)	Height (m)	Systolic blood pressure	Diastolic blood pressure	Tempe-rature (°F)	Diabetes
P. Lee	35	Female	A Rh⁺	50	1.52	68	112	98.7	0
A. Patel	70	Female	O Rh⁻	41	1.55	76	125	98.6	0
W. Hands	77	Female	O Rh⁻	104	1.71	107	145	98.3	1
P. Rice	45	Female	O Rh⁺	64	1.74	101	132	98.6	0

Table B.5. Chapter 3, question 7 answer (male patients)

Name	Age	Gender	Blood group	Weight (kg)	Height (m)	Systolic blood pressure	Diastolic blood pressure	Temperature (°F)	Diabetes
R. Jones	52	Male	O Rh⁻	115	1.77	110	154	98.5	1
J. Smith	45	Male	O Rh⁺	96	1.83	88	136	98.8	0
M. Owen	24	Male	A Rh⁻	79	1.82	65	105	98.7	0
S. Green	43	Male	O Rh⁻	109	1.89	114	159	98.9	1
N. Cook	68	Male	A Rh⁺	73	1.76	108	136	99.0	0
F. Marsh	28	Male	O Rh⁺	136	1.78	121	165	98.7	1

Chapter 4

1. See Table B.6
2a. See Table B.7
2b. See Table B.8
2c. See Table B.9
3. See Figure B.1
4. See Figure B.2

Table B.6. Chapter 4, question 1 answer

		Store		
		New York, NY	Washington, DC	Totals
	Laptop	1	2	3
Product	**Printer**	2	2	4
category	**Scanner**	4	2	6
	Desktop	3	2	5
	Totals	10	8	18

Table B.7. Chapter 4, question 2a answer

Customer	Number of observations	Sum of sales price ($)
B. March	3	1700
J. Bain	1	500
T. Goss	2	750
L. Nye	2	900
S. Cann	1	600
E. Sims	1	700
P. Judd	2	900
G. Hinton	4	2150
H. Fu	1	450
H. Taylor	1	400

Table B.8. Chapter 4, question 2b answer

Store	Number of observations	Mean sale price ($)
New York, NY	10	485
Washington, DC	8	525

Table B.9. Chapter 4, question 2c answer

Product category	Number of observations	Sum of profit ($)
Laptop	3	470
Printer	4	360
Scanner	6	640
Desktop	5	295

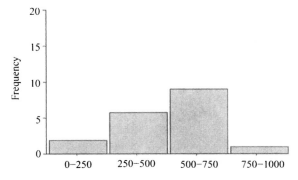

Figure B.1. Frequency distribution

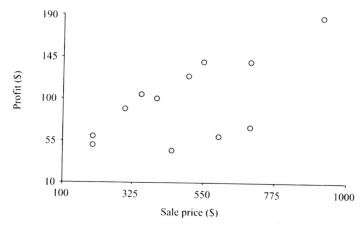

Figure B.2.   Scatterplot

Chapter 5

    1a. 45

    1b. 45

    1c. 48.7

    1d. 53

    1e. 324.9

    1f. 18.02

    1g. See Table B.10

    1h. 0.22

    1i. −1.5

    2. 24.14 − 25.86

    3. 0.386 − 0.534

Table B.10.   Chapter 5, question 1g answer

Name	Age	z-score Age
P. Lee	35	−0.76
R. Jones	52	0.18
J. Smith	45	−0.21
A. Patel	70	1.18
M. Owen	24	−1.37
S. Green	43	−0.32
N. Cook	68	1.07
W. Hands	77	1.57
P. Rice	45	−0.21
F. Marsh	28	−1.15

4a. H_o: $\mu = 2$

4b. H_a: $\mu < 2$ where μ is the average call connection time.

4c. –2.72

4d. 0.0033

4e. The phone company can make the claim.

5a. H_o: $\pi = 0.9$, H_a: $\pi > 0.9$ where π is the proportion of customers pleased with the service of the bank.

5b. 0.33

5c. 0.3707

5d. The bank cannot make the claim.

6a. H_O: $\mu_1 = \mu_2$, H_a: $\mu_1 > \mu_2$ where μ_1 is the average tomato plant height grown with fertilizer X and μ_2 is the average tomato plant height grown with fertilizer Y.

6b. 2.82

6c. 0.0024

6d. The company can make the claim.

7a. H_0: $\pi_1 = \pi_2$, H_a: $\pi_1 < \pi_2$ where π_1 is the proportion of defects, using manufacturer A, and π_2 is the proportion of defects, using manufacturer B.

7b. 0.54

7c. 0.2946

7d. The company cannot make the claim.

8a. H_O:$\mu_D = 0$, H_a:$\mu_D \neq 0$ where μ_D is the difference between the wear of the gloves.

8b. 15.36

8c. Practically zero.

8d. Yes.

9a. H_0: There is no relationship, H_a: There is a relationship.

9b. 2.18

9c. Cannot make the claim.

10a. H_0: There is no difference, H_a: There is a difference.

10b. 1.48

10c. Cannot make the claim.

11. 0.98

Chapter 6

1a. 4.8

1b. 2.8

1c. 0

Table B.11.   Chapter 7, question 2b answer

Observation	Actual	Predicted	Residual
1	13.7	12.4	1.3
2	17.5	16.1	1.4
3	8.4	6.7	1.7
4	16.2	15.7	0.5
5	5.6	8.4	−2.8
6	20.4	15.6	4.8
7	12.7	13.5	−0.8
8	5.9	6.4	−0.5
9	18.5	15.4	3.1
10	17.2	14.5	2.7
11	5.9	5.1	0.8
12	9.4	10.2	−0.8
13	14.8	12.5	2.3
14	5.8	5.4	0.4
15	12.5	13.6	−1.1
16	10.4	11.8	−1.4
17	8.9	7.2	1.7
18	12.5	11.2	1.3
19	18.5	17.4	1.1
20	11.7	12.5	−0.8

   2. 2.24

   3. Support $= 0.47$, Confidence $= 1$, Lift $= 1.89$

4a. 0.73

4b. 1.0

Chapter 7

1a. 0.85

1b. 0.15

1c. 0.89

1d. 0.82

2a. 0.87

2b. See Table B.11

3a. Height $= -0.071 + 0.074$ Fertilizer

3b. 0.98

   4. $352,600

   5. Brand B

   6. 0.56

Glossary

Accuracy. The accuracy reflects the number of times the model is correct.

Activation function. This is used within a neural network to transform the input level into an output signal.

Aggregation. A process where the data is presented in a summary form, such as average.

Alternative hypothesis. Within a hypothesis test, the alternative hypothesis (or research hypothesis) states specific values of the population that are possible when the null hypothesis is rejected.

Antecedent. An antecedent is the statement or statements in the IF-part of a rule.

Applying predictive models. Once a predictive model has been built, the model can be used or applied to a data set to predict a response variable.

Artificial neural network. See neural network.

Associative rules. Associative rules (or association rules) result from data mining and present information in the form "if X then Y".

Average. See mean.

Average linkage. Average linkage is the average distance between two clusters.

Backpropagation. A method for training a neural network by adjusting the weights using errors between the current prediction and the training set.

Bin. Usually created at the data preparation step, a variable is often broken up into a series of ranges or bins.

Binary variable. A variable with two possible outcomes: true (1) or false (0).

Binning. Process of breaking up a variable into a series of ranges.

Box plot. Also called a box-and-whisker plot, it is a way of graphically showing the median, quartiles and extreme values, along with the mean.

Box-Cox transformation. Often used to convert a variable to a normal distribution.

Building predictive models. This is the process of using a training set of examples and creating a model that can be used for prediction.

Business analyst. A business analyst specializes in understanding business needs and required solutions.

Categorical data. Data whose values fall into a finite number of categories.

Central limit theorem. States that the distribution of mean values will increasingly follow a normal distribution as the number of observations increases.

Chi-square. The chi-square statistic is often used for analyzing categorical data.

Churn. Reflects the tendency of subscribers to switch services.

Classification and Regression Trees (CART). Decision trees used to generate predictions.

Making Sense of Data: A Practical Guide to Exploratory Data Analysis and Data Mining,
By Glenn J. Myatt
Copyright © 2007 John Wiley & Sons, Inc.

Classification model. A model where the response variable is categorical.

Classification tree. A decision tree that is used for prediction of categorical data.

Cleaning (data). Data cleaning refers to the detecting and correcting of errors in the data preparation step.

Cleansing. See cleaning.

Clustering. Clustering attempts to identify groups of observations with similar characteristics.

Complete linkage. Maximum distance between an observation in one cluster and an observation in another one.

Concordance. Reflects the agreement between the predicted and the actual response.

Confidence interval. An interval used to estimate a population parameter.

Confidence level. A probability value that a confidence interval contains the population parameter.

Constant. A column of data where all values are the same.

Consumer. A consumer is defined in this context as one or more individuals who will make use of the analysis results.

Contingency table. A table of counts for two categorical variables.

Continuous variable. A continuous variable can take any real number within a range.

Correlation coefficient (r). A measure to determine how closely a scatterplot of two continuous variables falls on a straight line.

Cross validation. A method for assessing the accuracy of a regression or classification model. A data set is divided up into a series of test and training sets, and a model is built with each of the training set and is tested with the separate test set.

Customer Relationship Management (CRM). A database system containing information on interactions with customers.

Data. Numeric information or facts collected through surveys or polls, measurements or observations that need to be effectively organized for decision making.

Data analysis. Refers to the process of organizing, summarizing and visualizing data in order to draw conclusions and make decisions.

Data matrix. See data table.

Data mining. Refers to the process of identifying nontrivial facts, patterns and relationships from large databases. The databases have often been put together for a different purpose from the data mining exercise.

Data preparation. Refers to the process of characterizing, cleaning, transforming, and subsetting data prior to any analysis.

Data table. A table of data where the rows represent observations and the columns represent variables.

Data visualization. Refers to the presentation of information graphically in order to quickly identify key facts, trends, and relationships in the data.

Data warehouse. Central repository holding cleaned and transformed information needed by an organization to make decisions, usually extracted from an operational database.

Decimal scaling. Normalization process where the data is transformed by moving the decimal place.

Decision tree. A representation of a hierarchical set of rules that lead to sets of observations based on the class or value of the response variable.

Deployment. The process whereby the results of the data analysis or data mining are provided to the user of the information.

Descriptive statistics. Statistics that characterize the central tendency, variability, and shape of a variable.

Dichotomous variable. A variable that can have only two values.

Discrete variable. A variable that can take only a finite number of values.

Discretization. A process for transforming continuous values into a finite set of discrete values.

Dummy variable. Encodes a particular group of observations where 1 represents its presence and 0 its absence.

Embedded data mining. An implementation of data mining into an existing database system for delivery of information.

Entropy. A measurement of the disorder of a data set.

Error rate. Reflects the number of times the model is incorrect.

Euclidean distance. A measure of the distance between two points in n-dimensional space.

Experiment. A test performed under controlled conditions to test a specific hypothesis.

Exploratory data analysis. Processes and methods for exploring patterns and trends in the data that are not known prior to the analysis. It makes heavy use of graphs, tables, and statistics.

Feed-forward. In neural networks, feed-forward describes the process where information is fed through the network from the input to the output layer.

Frequency distribution. Description of the number of observations for items or consecutive ranges within a variable.

Frequency polygram. A figure consisting of lines reflecting the frequency distribution.

Gain. Measures how well a particular splitting of a decision tree separates the observations into specific classes.

Gaussian distribution. See normal distribution.

Gini. A measure of disorder reduction.

Graphs. An illustration showing the relationship between certain quantities.

Grouping. Methods for bringing together observations that share common characteristics.

Hidden layer. Used in neural networks, hidden layers are layers of nodes that are placed between the input and output layers.

Hierarchical agglomerative clustering. A bottom-up method of grouping observations creating a hierarchical classification.

Histogram. A graph showing a variable's discrete values or ranges of values on the x-axis and counts or percentages on the y-axis. The number of observations for each value or range is presented as a vertical rectangle whose length is proportionate to the number of observations.

Holdout. A series of observations that are set aside and not used in generating any predictive model but that are used to test the accuracy of the models generated.

Hypothesis test. Statistical process for rejecting or not rejecting a claim using a data set.

Inferential statistics. Methods that draw conclusions from data.

Information overload. Phenomena related to the inability to absorb and manage effectively large amounts of information, creating inefficiencies, stress, and frustration. It has been exacerbated by advances in the generation, storage, and electronic communication of information.

Input layer. In a neural network, an input layer is a layer of nodes, each one corresponding to a set of input descriptor variables.

Intercept. Within a regression equation, the point on the y-axis where x is 0.

Interquartile range. The difference between the first and third quartile of a variable.

Interval scale. A scale where the order of the values has meaning and where the difference between pairs of values can be meaningfully compared. The zero point is arbitrary.

Jaccard distance. Measures the distance between two binary variables.

K-means clustering. A top-down grouping method where the number of clusters is defined prior to grouping.

K-nearest neighbors (kNN). A prediction method, which uses a function of the k most similar observations from the training set to generate a prediction, such as the mean.

Kurtosis. Measure that indicates whether a variable's frequency distribution is peaked or flat compared to a normal distribution.

Leaf. A node in a tree or network with no children.

Learning. A process whereby a training set of examples is used to generate a model that understands and generalizes the relationship between the descriptor variables and one or more response variables.

Least squares. A common method of estimating weights in a regression equation that minimizes the sum of the squared deviation of the predicted response values from the observed response values.

Linear relationship. A relationship between variables that can be expressed as a straight line if the points are plotted in a scatterplot.

Linear regression. A regression model that uses the equation for a straight line.

Linkage rules. Alternative approaches for determining the distance between two clusters.

Logistic regression. A regression equation used to predict a binary variable.

Mathematical models. The identification and selection of important descriptor variables to be used within an equation or process that can generate useful predictions.

Mean. The sum of all values in a variable divided by the number of values.

Medium. The value in the middle of a collection of observations.

Min–max normalization. Normalizing a variable value to a predetermine range.

Missing data. Observations where one or more variables contain no value.

Mode. The most commonly occurring value in a variable.

Models. See mathematical model.

Nominal scale. A scale defining a variable where the individual values are categories and no inference can be made concerning the order of the values.

Multilinear regression. A linear regression equation comprising of more than one descriptor variable.

Multiple regression. A regression involving multiple descriptor variables.

Negative relationship. A relationship between variables where one variable increases while the other variable decreases.

Neural network. A nonlinear modeling technique comprising of a series of interconnected nodes with weights, which are adjusted as the network learns.

Node. A decision point within a decision tree and a point at which connections join within a neural network.

Nominal scale. A variable is defined as being measured on a nominal scale if the values cannot be ordered.

Nonhierarchical clustering. A grouping method that generates a fixed set of clusters, with no hierarchical relationship quantified between the groups.

Nonlinear relationship. A relationship where while one or more variables increase the change in the response is not proportional to the change in the descriptor(s).

Nonparametric. A statistical procedure that does not require a normal distribution of the data.

Normal distribution. A frequency distribution for a continuous variable, which exhibits a bell-shaped curve.

Normalizations (standardization). Mathematical transformations to generate a new set of values that map onto a different range.

Null hypothesis. A statement that we wish to clarify by using the data.

Observation. Individual record in a data table.

Observational study. A study where the data collected was not randomly obtained.

Occam's Razor. A general rule to favor the simplest theory to explain an event.

On-Line Analytical Processing (OLAP). Tools that provide different ways of summarizing multidimensional data.

Operational database. A database containing a company's up-to-date and modifiable information.

Ordinal scale. A scale measuring a variable that is made of items where the order of the items has meaning.

Outlier. A value that lies outside the boundaries of the majority of the data.

Output layer. A series of nodes in a neural network that interface with the output response variables.

Overfitting. This is when a predictive model is trained to a point where it is unable to generalize outside the training set of examples it was built from.

Paired test. A statistical hypothesis test used when the items match and the difference is important.

Parameter. A numeric property concerning an entire population.

Parametric. A statistical procedure that makes assumptions concerning the frequency distributions.

Placebo. A treatment that has no effect, such as a sugar pill.

Point estimate. A specific numeric estimate of a population parameter.

Poll. A survey of the public.

Population. The entire collection of items under consideration.

Positive relationship. A relationship between variables where as one variable increases the other also increases.

Prediction. The assignment using a prediction model of a value to an unknown field.

Predictive model (or prediction model). See mathematical model.

Predictor. A descriptor variable that is used to build a prediction model.

p-value. A *p-value* is the probability of obtaining a result at least as extreme as the null hypothesis.

Range. The difference between the highest and the lowest value.

Ratio scale. A scale where the order of the values and the differences between values has meaning and the zero point is nonarbitrary.

Regression trees. A decision tree used to predict a continuous variable.

Residual. The difference between the actual data point and the predicted data point.

Response variable. A variable that will be predicted using a model.

r-squared. A measure that indicates how well a model predicts.

Sample. A set of data selected from the population.

Sampling error. Error resulting from the collection of different random samples.

Sampling distribution. Distribution of sample means.

Scatterplot. A graph showing two variables where the points on the graph correspond to the values.

Segmentation. The process where a data set is divided into separate data tables, each sharing some common characteristic.

Sensitivity. Reflects the number of correctly assigned positive values.

Similarity. Refers to the degree two observations share common or close characteristics.

Simple linear regression. A regression equation with a single descriptor variable mapping to a single response variable.

Simple nonlinear regression. A regression equation with a single descriptor variable mapping to a single response variable where whenever the descriptor variable increases, the change in the response variable is not proportionate.

Simple regression. A regression model involving a single descriptor variable.

Single linkage. Minimum distance between an observation in one cluster and an observation in another.

Skewness. For a particular variable, skewness is a measure of the lack of symmetry.

Slope. Within a simple linear regression equation, the slope reflects the gradient of the straight line.

Specificity. Reflects the number of correctly assigned negative values.

Splitting criteria. Splitting criteria are used within decision trees and describe the variable and condition in which the split occurred.

Spreadsheet. A software program to display and manipulate tabular data.

Standard deviation. A commonly used measure that defines the variation in a data set.

Standard error of the mean. Standard deviation of the means from a set of samples.

Standard error of the proportion. Standard deviation of proportions from a set of samples.

Statistics. Numeric information calculated on sample data.

Subject matter expert. An expert on the subject of the area on which the data analysis or mining exercise is focused.

Subset. A portion of the data.

Sum of squares of error (SSE). This statistic measures the total deviation of the response from the predicted value.

Summary table. A summary table presents a grouping of the data where each row represent a group and each column details summary information, such as counts or averages.

Supervised learning. Methods, which use a response variable to guide the analysis.

Support. Represents a count or proportion of observations within a particular group included in a data set.

Survey. A collection of questions directed at an unbiased random section of the population, using nonleading questions.

Temporal data mining. See time-series data mining.

Test set. A set of observations that are not used in building a prediction model, but are used in testing the accuracy of a prediction model.

Textual data mining. The process of extracting nontrivial facts, patterns, and relationships from unstructured textual documents.

Time-series data mining. A prediction model or other method that uses historical information to predict future events.

Training set. A set of observations that are used in creating a prediction model.

Transforming (data). A process involving mathematical operations to generate new variables to be used in the analysis.

Two-sided hypothesis test. A hypothesis test where the alternative hypothesis population parameter may lie on either side of the null hypothesis value.

Type I error. Within a hypothesis test, a type I error is the error of incorrectly rejecting a null hypothesis when it is true.

Type II error. Within a hypothesis test, a type II error is the error of incorrectly not rejecting a null hypothesis when it should be rejected.

Unsupervised learning. Analysis methods that do not use any data to guide the technique operations.

Value mapping. The process of converting into numbers variables that have been assigned as ordinal and described using text values.

Variable. A defined quantity that varies.

Variance. The variance reflects the amount of variation in a set of observations.

Venn Diagram. An illustration of the relationship among and between sets.

z-score. The measure of the distance in standard deviations of an observation from the mean.

Bibliography

A Guide To The Project Management Body Of Knowledge (PMBOK Guides) Third Edition, Project Management Institute, Pennsylvania, 2004

AGRESTI, A., *Categorical Data Analysis Second Edition*, Wiley-Interscience, 2002

ALRECK, P. L., and R. B. SETTLE, *The Survey Research Handbook Third Edition*, McGraw-Hill/Irwin, Chicago, 2004

ANTONY, J., *Design of Experiments for Engineers and Scientists*, Butterworth-Heinemann, Oxford, 2003

BARRENTINE, L. B., *An Introduction to Design of Experiments: A Simplified Approach*, ASQ Quality Press, Milwaukee, 1999

BERKUN, S., *The Art of Project Management*, O'Reily Media Inc., Sebastopol, 2005

BERRY, M. W., *Survey of Text Mining: Clustering, Classification, and Retrieval*, Springer-Verlag, New York, 2003

BERRY, M. J. A., and G. S. LINDOFF, *Data Mining Techniques for Marketing, Sales and Customer Support Second Edition*, John Wiley & Sons, 2004

COCHRAN, W. G., and G. M. COX, *Experimental Designs Second Edition*, John Wiley & Sons Inc., 1957

CRISTIANINI, N., and J. SHAWE-TAYLOR, *An Introduction to Support Vector Machines and Other Kernel-Based Learning Methods*, Cambridge University Press, 2000

DASU, T., and T. JOHNSON, *Exploratory Data Mining and Data Cleaning*, John Wiley & Sons Inc., Hoboken, 2003

DONNELLY, R. A., *Complete Idiot's Guide to Statistics*, Alpha Books, New York, 2004

EVERITT, B. S., S. LANDAU, and M. LEESE, *Cluster Analysis Fourth Edition*, Arnold, London, 2001

HAN J., and M. KAMBER, *Data Mining: Concepts and Techniques Second Edition*, Morgan Kaufmann Publishers, 2005

HAND, D. J., H. MANNILA, and P. SMYTH, *Principles of Data Mining*, Morgan Kaufmann Publishers, 2001

HASTIE, T., R. TIBSHIRANI, and J. H. FRIEDMAN, *The Elements of Statistical Learning*, Springer-Verlag, New York, 2003

FAUSETT, L. V., *Fundamentals of Neural Networks*, Prentice Hall, New York, 1994

FOWLER, F. J., *Survey Research Methods (Applied Social Research Methods) Third Edition*, SAGE Publications Inc., Thousand Oaks, 2002

FREEDMAN, D., R. PISANI, and R. PURVES, *Statistics Third Edition*, W. W. Norton, New York, 1997

GUIDICI, P., *Applied Data Mining: Statistical Methods for Business and Industry*, John Wiley & Sons Ltd., 2005

KACHIGAN, S. K., *Multivariate Statistical Analysis: A Conceptual Introduction*, Radius Press, New York, 1991

KERZNER, H., *Project Management: A Systems Approach to Planning, Scheduling and Controlling Ninth Edition*, John Wiley & Sons, 2006

KIMBALL, R., and M. ROSS, *The Data Warehouse Toolkit: The Complete Guide to Dimensional Modeling Second Edition*, Wiley Publishing Inc., Indianapolis, 2002

KLEINBAUM, D. G., L. L. KUPPER, K. E. MULLER, and A. NIZAM, *Applied Regression Analysis and Other Multivariate Methods Third Edition*, Duxbury Press, 1998

KWOK, S., and C. CARTER, Multiple Decision Trees, In: SCHACHTER, R. D., T. S. LEVITT, L. N. KANAL, and J. F. LEMER (eds), *Artificial Intelligence 4*, pp. 327–335, Elsevier Science, Amsterdam, 1990

JOLLIFFE, I. T., *Principal Component Analysis Second Edition*, Springer-Verlag, New York, 2002

JACKSON, J. E., *A User's Guide to Principal Components*, John Wiley & Sons, Inc., Hoboken, 2003

Making Sense of Data: A Practical Guide to Exploratory Data Analysis and Data Mining,
By Glenn J. Myatt
Copyright © 2007 John Wiley & Sons, Inc.

LAST, M., A. KANDEL, and H. BUNKE, *Data Mining in Time Series Databases*, World Scientific Publishing Co. Pte. Ltd., 2004

LEVINE, D. M., and D. F. STEPHAN, *Even You Can Learn Statistics: A guide for everyone who has ever been afraid of statistics*, Pearson Education Inc., Upper Saddle River, 2005

MONTGOMERY, D. C., *Design and Analysis of Experiments Sixth Edition*, John Wiley & Sons Inc., Noboken, 2005

NEWMAN, D.J., S. HETTICH, C.L. BLAKE, and C.J. MERZ, *UCI Repository of machine learning databases* [http://www.ics.uci.edu/~mlearn/MLRepository.html], University of California, Department of Information and Computer Science, Irvine, CA, 1998

OPPEL, A., *Databases Demystified: A Self-Teaching Guide*, McGraw-Hill/Osborne, Emeryville, 2004

PEARSON, R. K., *Mining Imperfect Data: Dealing with Contamination and Incomplete Records*, Society of Industrial and Applier Mathematics, 2005

PYLE, D., *Data Preparation and Data Mining*, Morgan Kaufmann Publishers Inc., San Francisco, 1999

QUINLAN, J. R., *C4.5: Programs for Machine Learning*, Morgan Kaufmann Publishers Inc., San Mateo, 1993

REA, L. M., and R. A. PARKER, *Designing and Conducting Survey Research: A Comprehensive Guide Third Edition*, Jossey Bass, San Francisco, 2005

RUDD, O. P., *Data Mining Cookbook*, John Wiley & Sons, 2001

RUMSEY, D., *Statistics for Dummies*, Wiley Publishing Inc., Hoboken, 2003

TANG, Z., and J. MACLENNAN, *Data Mining with SQL Server 2005*, Wiley Publishing Inc., Indianapolis, 2005

TUFTE, E. R., *Envisioning Information*, Graphics Press, Cheshire, 1990

TUFTE, E. R., *Visual Explanation: Images and Quantities, Evidence and Narrative*, Graphics Press, Cheshire, 1997

TUFTE, E. R., *The Visual Display of Qualitative Information Second Edition*, Graphics Press, Cheshire, 2001

WEISS, S. M., N. INDURKHYA, T. ZHANG, and F. J. DAMERAU, *Text Mining: Predictive Methods for Analyzing Unstructured Information*, Springer-Verlag, New York, 2004

WITTEN, I. H, and E. FRANK, *Data Mining: Practical Machine Learning Tools and Techniques with Java Implementations*, Morgan Kaufmann Publishers, 2000

WOLD, H., Soft modeling by latent variables: the non-linear iterative partial least squares (NIPALS) approach. *Perspectives in Probability and Statistics (papers in honour of M. S. Bartlett on the occasion of his 65th birthday)*, pp. 117–142, Applied Probability Trust, University of Sheffield, Sheffield, 1975

Index

Making Sense of Data: A Practical Guide to Exploratory Data Analysis and Data Mining,
By Glenn J. Myatt
Copyright © 2007 John Wiley & Sons, Inc

Breinigsville, PA USA
02 February 2010
231795BV00002B/1/P